PHYSICS RESEARCH AND TECHNOLOGY

THE INFINITE HISTORY OF NOW

A TIMELESS BACKGROUND FOR CONTEMPORARY PHYSICS

PHYSICS RESEARCH AND TECHNOLOGY

Additional books in this series can be found on Nova's website under the Series tab.

Additional e-books in this series can be found on Nova's website under the e-book tab.

PHYSICS RESEARCH AND TECHNOLOGY

THE INFINITE HISTORY OF NOW

A TIMELESS BACKGROUND FOR CONTEMPORARY PHYSICS

DAVIDE FISCALETTI

AND

AMRIT SRECKO SORLI

nova publishers

New York

For permission to use material from this book please contact us:
Telephone 631-231-7269; Fax 631-231-8175
Web Site: http://www.novapublishers.com

Library of Congress Cataloging-in-Publication Data

ISBN: 978-1-63117-283-0

Published by Nova Science Publishers, Inc. † New York

CONTENTS

Preface		**ix**
Introduction		**xi**
	References	*xvi*
Chapter 1	**Timeless Universe**	**1**
	Abstract	*1*
	1.1. Time as a Measuring System of the Numerical Order of Material Motion	*2*
	1.2. Planck Time as a Fundamental Unit of Numerical Order. Physical Time and Mathematical Time	*8*
	1.3. Special Relativity in a Three-Dimensional Euclid Space	*10*
	1.4. Time as Numerical Order Resolves Zeno Problems on Motion	*14*
	1.5. Time as Numerical Order Resolves Hypothetical Travel into Past and Twin Paradox	*15*
	1.6. Psychological Time Creates Temporal Experience "Past-Present-Future"	*15*
	1.7. EPR Experiment and the Idea of Space as an Immediate Information Medium	*16*
	1.8. Bohm's Quantum Potential and Its Link with the Idea of Space as an Immediate Information Medium	*18*

	1.9. The Problem of Time-Symmetry. The Immediate Symmetric Interpretation of Quantum Non-Locality	*23*
	1.10. Immediate and Non-Immediate Physical Phenomena	*29*
	1.11. Immediate Physical Phenomena and Their Explanation Through the Idea of Space as an Immediate Information Medium	*31*
	1.12. A Three-Dimensional Timeless Quantum Vacuum as a Fundamental Bridge Between Gravitation and the Quantum Behaviour	*35*
	1.13. Relativistic Effects of Relative Velocity of Material Change Start with Massive Particles	*51*
	1.14. Timeless Universe in Dynamic Equilibrium	*53*
	References	*56*
Chapter 2	**Towards a Unified View of the Physical World**	**61**
	Abstract	*61*
	2.1. Towards a Deeper Structure of Subquantum Physics as a Fundamental Medium of the Universe	*62*
	2.2. About the Quantum Gravity Domain and Unifying Theories: Strings, Loops ... and the E8 Theory	*66*
	2.3. Modelling Physical Reality at the Fundamental Level: Being Versus Becoming and Atomism Versus Monism	*73*
	2.4. The Postulates of the Theory of Everything	*80*
	2.5. The Interpretation of Subatomic Particles in the Timeless Space	*85*
	2.6. Quantum Potential as "Special State of the Timeless Space in the Presence of Microscopic Processes"	*89*
	2.7. The Double-Slit Interference	*92*
	2.8. Tunnelling	*94*
	2.9. The Fundamental Interactions in the Timeless Space	*96*
	2.10. About the Wave Picture of the Timeless Theory of Everything	*98*

2.11. A New Interpretation of Superstring Theory and of Loop Quantum Gravity in the Timeless Space *102*
2.12. Towards a Holistic View of the Physical World *105*
References *108*

Index **111**

PREFACE

The development of theoretical physics can be seen as a continuous improvement of the models of space and time. In particular, the beginning of the 21st century brought a new vision into quantum physics: quantum particles move in space only, time measured with clocks is merely a numerical order of particles' motion. This so called "a-temporal" view of physics offers a radically new understanding of the micro and macro universe. On the basis of the notion of time as numerical order of material change, in the book "The infinite history of now. A timeless background for contemporary physics", the authors Davide Fiscaletti and Amrit Sorli introduce suggestive unitary and holistic perspectives in the interpretation and explanation of several physical phenomena of contemporary physics. It is shown that the concept of time, as a mathematical quantity measuring the numerical order of material changes, resolves Zeno problems on motion, hypothetical travel into past, twin paradox of special relativity and allows new perspectives to be opened in the interpretation and explanation of Einstein-Podolski-Rosen experiment and of non-local correlations between subatomic particles. Moreover, a timeless description of the gravitational interaction and the idea of a timeless three-dimensional quantum vacuum as a fundamental bridge between gravity and quantum behaviour of matter are developed. Finally, in the second part of the book, after analysing the fundamental features of some relevant unitary theories of contemporary physics (such as superstring theory and loop quantum gravity) and making some considerations about being versus becoming and atomism versus monism in modelling physical reality at a fundamental level, it is shown that the replacement of the concept of time with the numerical order of material change provides a new suggestive interpretation of the fundamental unitary field theories opening the doors to a real “timeless theory of everything”. The use of a simple and fluid language makes the book accessible to a wide reading public.

INTRODUCTION

The beginning of the 21st century brought a new vision into quantum physics: quantum particles move in space only, time that we measure with clocks is merely a numerical order of particles' motion. Past-present-future is merely a mind model into which physicist experiences material world. This so called "a-temporal" view of physics offers a radically new understanding of the micro and macro universe: the world in which we live is a non-created dynamic system in a permanent equilibrium. All worries about "creation" and "duration" of the universe are forgotten now. In this book our main task is to discover the timeless laws according to which this divine existence works.

The whole development of theoretical physics can be seen as a continuous improvement of the models of space and time. No physical law can be formulated without being collocated in an opportune space-temporal arena. This fact stimulated many times in the history of physics the idea of a tight connection between the physical processes and the global arena in which they take place. This connection must be intended in a realistic sense, and implies a series of radical assumptions as regards the fundamental sub-structure and the nature of space and time which delineate a series of deep perspective changes about our way of working with the physical theories in the research of an unitary view of the world.

Newtonian absolute conception has been the first attempt to ascribe a fundamental role to space and time in order to explain the properties of matter. In Newtonian physics space and time are absolute entities and define a stage over which the action of physics takes place, the various dynamical entities being the actors. In the 20th century three fundamental theories have changed the understanding and explanation of the phenomena occurring in the universe: special relativity, general relativity and quantum theory. Einstein's special relativity (which studies phenomena occurring at speeds comparable to the light speed, inside inertial systems) has determined the first important changes in the Newtonian picture. In special relativity space and time are intrinsically linked, united into a manifold called "space-time", a smooth four-dimensional continuum. Despite time is no more absolute and the idea of absolute simultaneity must be abandoned (which really shatter some Newtonian ideas), however, just as in Newtonian physics, also in special relativity there is still a fixed space-time which serves as the arena for all of physics. It is the stage on which the drama of evolution unfolds. Actors are particles and fields. The stage constrains what the actors can do. But the actors cannot influence the stage; Minkowskian geometry characteristic of special relativity is immune from change. The Newtonian construct of a fixed background stage over which physics happens is not altered by special relativity.

A more profound change in the picture of the world happened with Einstein's general relativity. General relativity is the universally accepted theory that allows us to describe the gravitational interaction. The central discovery of general relativity is that Newton's space-time and the gravitational field are the same thing and that the dynamics of the gravitational field and of any other dynamical object is fully relational, in the Aristotelian-Cartesian sense. Matter curves space-time. The space-time metric is no longer fixed. This is a crucial paradigm shift: one has to abandon the idea that space-time is fixed, immune to change, one has to encode gravity into the very geometry of space-time, thereby making this geometry dynamical. With general relativity Einstein abandoned the tenet that geometry is inert and made it a physical entity that interacts with matter. As a consequence, general relativity treats time in a peculiar way, as compared to pre-general relativistic physics. The absolute time of Newton has disappeared. In its place, there are different possible internal times, related to specific physical clocks. Only the evolution with respect to these physical clocks makes sense. The way evolution is treated in general relativity is therefore more subtle than in pre-general relativistic physics. Change is not described as evolution of physical variables as a function of a preferred independent observable time variable. Instead, it is described in terms of a relation among equal footing variables. In general relativity, there is not a preferred and observable quantity that plays the role of independent parameter of the evolution, as there is in non-relativistic mechanics. General relativity describes the relative evolution of observable quantities, not the evolution of quantities as functions of a preferred one.

On the other hand, another relevant 20th century scientific revolution happened with quantum theory. Quantum mechanics is perhaps the 20th century physical theory that has contributed most profoundly to change our understanding of the universe and at the same time has created the most complex interpretative problems as regards what it says about the world. Indeed quantum mechanics introduces much more spread scenarios and perspectives than those offered by every previous physical theory. Much of 20th century physics can be seen as the triumph exploration of the new scenarios opened by quantum mechanics. Since its birth – happened in the second half of the 20s of the last century – quantum mechanics revealed itself as the essential tool for the investigation of the physical world, leading to the development of atomic physics, nuclear physics, particle physics, condensed matter physics, quantum cosmology, etc....

In general terms, we can define quantum mechanics as a reversible extension of classical mechanics in such a way to take into account the properties of atomic and subatomic systems, the microscopic aspects of the interaction between radiation and matter. In the analysis of these phenomena we need to introduce a new fundamental constant of nature, namely the Planck constant h[1]. The quantum theory determined many changes in the description of the world by showing in particular that in microphysics the different physical quantities (for example, energy, momentum) have not a continuous spectrum, but can assume only a discrete set of values, namely are quantized.

Now, at the beginning of the new millennium, certain conceptual confusions in different physical theories, models and ideas are evident. The most important problems lay in the fact that, on one hand, the two relativity theories and, on the other hand, quantum mechanics – despite their relevant successes from the predictive point of view – have created a picture of the physical world which is somewhat unclear, incomplete and fragmented. In fact, on one hand, the pictures of reality that are determined by special relativity and general relativity (and that describe the space-temporal arena, in absence and in presence of gravity respectively) are precise and geometric pictures. On the other hand, the picture of reality that is offered by quantum theory (in the study of atomic and subatomic processes) turns out to be intrinsically probabilistic and indeterministic. The strategy to utilize two different pictures of the world in order to study two different domains of physical reality is clearly very little satisfactory from a conceptual point of view. The tendency to unification which always characterizes the physical research would require having one picture of the world in order to describe the different domains of nature.

Moreover, it is important to underline that Einstein's general theory of relativity and quantum theory – which perhaps can be considered the two greatest achievements of 20th century physics – are indeed mutually incompatible. Thus

[1] As is known, the Planck constant turns out to assume an important role in the microscopic world; it assumes instead a negligible value in the domain of classical mechanics.

our basic understanding of nature is not only incomplete and fragmented – but inconsistent. One clearly needs a new consistent theory describing the so-called "quantum gravity domain" which provides a unitary picture of the universe, incorporates the principles of both quantum mechanics and general relativity, and reduces to them in appropriate limits. Finding a new synthesis and a unitary picture of the universe represents therefore the most important challenge in today's fundamental physics.

What is matter? What is time? What is the meaning of "being somewhere"? What is the meaning of "now"? What is the meaning of "moving"? These are fundamental questions that we must face in order to find a novel picture of the world and to find a new synthesis in physics. We need to find a new answer to these questions – different from Newton's answer – which took into account what we have learned about the world with quantum mechanics and general relativity.

In order to find a novel picture of the world, the starting-point perspective of the authors of this book is that space-time is not the fundamental physical arena of the universe. There is no experimental evidence whatsoever to support the view that space-time exists as the basis of a fundamental physical reality. Space-time cannot be considered a primary physical structure, independent from matter. Already Einstein in his book *Relativity: The Special and General Theory* suggested that it makes no sense to think about space-time as an independent physical entity. The basic medium into which material change occurs is the gravitational field, namely a "space-matter structure", in other words a gravitational space. Space-time can be thus seen only as a structural quality of a gravitational field: "Space-time does not claim existence on its own but only as a structural quality of the [gravitational] field" [1].

Our experience of space is based on observations of material objects. Our experience of time derives from the changes and movements of matter into the universe. Time cannot be considered as a primary physical reality, as a quantity that is independent from the rest, as a quantity which flows on its own in the universe. Time is a mathematical coordinate which describes the motion of material objects. Time is a quantity associated with the stream of material changes and motions occurring in a timeless gravitational space and enters into existence when we measure it through clocks. Clocks run in a timeless gravitational space. With clocks we measure the numerical order of material changes that run into the universe.

As a consequence, space-time as a structural quality of the gravitational space can be considered an observer effect formed into the process of measurement. When there is no measurement, space-time exists only as a mathematical model and does not exist as a structural quality of the gravitational space.

Moreover, it is important to underline that in order to find a synthesis, and unification of general relativity and quantum mechanics, since the second half of the 20th century different strategies have been adopted. In particular, on one hand, according to string theory, the fundamental objects of nature are strings, one-dimensional objects characterized by different vibrations and these vibrations describe the different types of particles present in nature. String theory postulates that the origin, the basis of the laws of the universe is the vibratory character of nature. Taking into account the results of string theory, it seems thus legitimate to postulate that physical reality has a vibratory structure.

On the other hand, there are different physical theories that treat space as a quantum. In particular, loop quantum gravity suggests that at the most fundamental level physical space has a granular structure, namely is composed by elementary grains, a net of intersecting loops, having the Planck size[2]. According to the picture of the world predicted by loop quantum gravity all we observe in nature is not continuous but is quantized, has a discrete nature. In synthesis, loop quantum gravity predicts that the origin, the basis of the laws of the universe is the granular structure of space at the Planck scale. Taking into account the results of loop quantum gravity, it turns out to be permissible to assume that space has a granular structure at Planck scale. As a consequence, on the ground of the results of string theory and loop quantum gravity, it seems natural to expect that, at a fundamental level, physical reality has contemporarily a vibratory and granular nature.

In synthesis, according to the view proposed in this book, the timeless character, the vibratory and granular structure of physical universe can be considered as the most important perspectives of contemporary physics: time exists only as a coordinate associated with the stream of material change and the fundamental constituents of the universe are quanta of space having the size of the Planck length and vibrating at appropriate frequencies.

---

[2] The Planck length $l_p = \sqrt{\frac{\hbar G}{c^3}}$ (where $\hbar$ is Planck's reduced constant, G is the gravitational constant and c is the speed of light) is a fundamental unit of length equal to $1{,}31 \cdot 10^{-35}$ meters.

REFERENCES

[1] Einstein, A. *Relativity: the Special and General Theory*; Henry Holt: New York, NY, 1920; Bartleby.com, 2000, www.bartleby.com/173/.

Chapter 1

TIMELESS UNIVERSE

ABSTRACT

In Newtonian physics as well as in standard quantum mechanics, time is postulated as a special physical quantity that plays the role of the independent variable of physical evolution. Newton or Hamilton equations of classical mechanics, as well as the Schrödinger equation of quantum mechanics, are introduced on the basis of the underlying assumption that an idealized, absolute time exists in which the dynamics is defined. However, it is an elementary observation that we never really measure an absolute time that an idealized, absolute time never appears in laboratory measurements: we rather measure the frequency, speed and numerical order of material changes. What we realize in every experiment is to compare the motion of the physical system under consideration with the motion of a peculiar clock described by a peculiar thick. This means that the duration of material motions has not a primary physical existence, independent on the motion of matter.

Changes of state of the universe and, at the same time, changes of the state of any physical system can be considered the primary phenomena which generate evolution of the universe. This evolution can be described by introducing a mathematical parameter, which provides only the ordering of events. In the article *Projection evolution and delayed choice experiment*, the Polish physicists Gozdz and Stefanska showed that an evolution parameter, "numerical order", which provides only the order of events, can be easily introduced [1]. In the papers *Time is a reference system derived from light speed* and *Replacing time with numerical order of material change resolves Zeno problems on motion* published recently in "Physics Essays", the authors of this book have gone beyond by suggesting the following concept of time: the symbol of time appearing in all mathematical formalisms of physics is a number which represents the numerical order of material changes, i.e., motions and therefore only the numerical order of the motion of the system under consideration, which is obtained by the clock under consideration, exists [2, 3].

In this chapter we will analyse the view of a timeless universe where time exists only as a numerical order of material changes. The fundamental idea that is at the basis of the research of the authors is that evolution occurs in a three-dimensional space background, that a timeless three-dimensional space background corresponding to the observations and motion of material objects, is the fundamental arena of physical processes. We will see, in particular, that our view of a timeless three-dimensional arena, where time is only a mathematical quantity measuring the numerical order of material changes, resolves Zeno problems on motion, hypothetical travel into past, twin paradox of special relativity and allow new perspectives to be opened in the interpretation and explanation of Einstein-Podolski-Rosen experiment and of non-local correlations between subatomic particles. Finally, we will develop a timeless description of the gravitational interaction and we will suggest the idea of a timeless three-dimensional quantum vacuum as a fundamental bridge between gravity and quantum behaviour of matter and the cosmological view of a timeless universe in dynamic equilibrium.

1.1. Time as a Measuring System of the Numerical Order of Material Motion

Going deeper into the measurement of time we see that with clocks we measure numerical order $t_0, t_1, t_2 ... t_n$ of material change. On the basis of elementary perception, the passing of the time cannot be perceived directly as matter and space; we can perceive only the changes and movements of matter in space. Time cannot be considered a primary physical reality that flows on its own in the universe: it exists only as a quantity associated with the stream of material changes occurring in a timeless space. Time is only a coordinate of change (movement) and, therefore, when there is no change there is no time. Material change n is followed by material change $n+1$; material change $n+1$ is followed by material change $n+2$ and so on. Material change $n+1$ is "after" material change n equivalently as natural number 2 is "after" natural number 1. In other words, we suggest that the parameter time that we use in physics indeed represents only a quantity which indicates the numerical order of the event under consideration, just like the mathematical ensemble of natural numbers. Numerical order $t_0, t_1, t_2 ... t_n$ of material change runs in space only and not in time. There is no time existing behind numerical order of material change. "Past", "present" and "future" exist in space only as a numerical order of material change.

Clocks run in a timeless space. With the eyes we can perceive material change into the universe only, space-time into which this material change should happen is not perceivable. Space-time exists as a mathematical model into which one experiences universe. Phenomena run in a space-time manifold only in the mathematical models of reality, which sometimes become more real than reality itself, which instead – on the basis of our elementary perception – turns out to be timeless. The real stage in which natural phenomena happen is not space-time but is a timeless space where time exists only as a numerical order of material change. This is an alternative, different point of view from that conventionally adopted in physics, but is perhaps more correct and appropriate because is more coherent with experimental facts.

The results of several eminent authors suggest that the background space of physics is timeless and clock/time exists merely as a measuring system of the numerical order of physical events.

Already in 19th century the Austrian physicist Ernst Mach said: "It is utterly beyond our power to measure the changes of things by time. Quite the contrary, time is an abstraction, at which we arrive by means of the changes of things" [4]. Mach's view seems completely consistent with experimental facts: time does not exist as a primary reality and clocks provide only a numerical description of the stream of material changes and motions. The idea of a timeless universe was also discussed by Einstein and the German mathematician Gödel at the beginning of the '50 of the last century. They have been discussing the possibility that there is no time in the universe as science experiences it, that time into which material changes run exists only as a concept of our mind. In 1949, Gödel postulated also a theorem that states: "In any universe described by the theory of relativity, time cannot exist" [5].

Though the argument of time as coordinate of motion seems simple, elegant and epistemologically sound, it took nearly 100 years since the original quote by Mach before the English physicist Julian Barbour and the Italian physicist Bruno Bertotti were able to formulate this principle into a mathematically rigorous theory of time in 1982 [6]. The reason for this delay can not be attributed to technical complications, since the mathematics underlying the approach have been well understood since Jacobi[1], but rather to conceptual confusion surrounding how Mach's ideas are implemented in general relativity. Though it was clear that

Einstein was heavily influenced by Mach's ideas, general covariance (namely the invariance of physical laws under arbitrary change of reference frame) proved to be misleading as a way of implementing relationalism. However, through the papers of Barbour and his collaborators, we now have a clear picture of how Mach's ideas can be used to derive general relativity (see, for example, the references [7, 8]). From these works, we know that, in regards to time, Mach's view can be implemented in general relativity by using Jacobi's principle to determine the classical dynamics of the system. In Barbour's approach, a "time" intended as primary physical reality is eliminated and is replaced by an unphysical evolution parameter, a reparamaterization invariant quantity, called ephemeris time, which provides a measure of duration corresponding to the relative change in the positions of the particles in the system.

Today, the idea that time cannot be considered as a primary physical reality and the arena of physical processes is therefore timeless is receiving indeed more and more attention. For example, The English physicist Woodward argues that Mach's principle (according to which inertial motion is governed by the totality of masses in the universe) leads to the conclusion that time, as we normally treat it in our common experience and physical theory, is not a part of fundamental reality [9]. As it has been shown by Woodward, if inertia is relative and gravitationally induced, Mach's principle kills our conventional conception of time, makes difficult to avoid killing time. Also in the context of quantum mechanics, on the basis of the results of Woodward, the immediate interpretation of Mach's principle determines the weird consequences regarding timelessness: the price to pay in order to have clarity of understanding in quantum mechanics is to accept a radical timelessness. The Italian physicist of the university of Marseille Carlo Rovelli proposes the idea that an idealized time is not defined at the fundamental level (at the Planck scale), namely that, in the quantum gravity regime, time should be simply forgotten, that the concept of a absolute time, as used in Hamiltonian mechanics as well as in Schrödinger quantum mechanics, is not relevant in a fundamental description of quantum gravity [10-14]. Julian Barbour puts in evidence in *The Nature of Time*: "I will not claim that time can be definitely banished from physics; the universe might be infinite, and black holes present some problems for the time picture. Nevertheless, I think it is entirely possible, indeed likely, that time as such plays no role in the universe" [15].

Moreover, in the recent paper *Entropic dynamics, time and quantum theory*, the American physicist of the University of Albany-SUNY Ariel Caticha

[1] Carl Gustav Jacobi was a German mathematician who during the XIX century made important contributions in the study of differential equations developing a fundamental formulation of

proposed an "entropic dynamics" in which time is seen as a device to keep track of change [16]. Caticha's model shows how a dynamics driven by entropy naturally leads to an "entropic" notion of time. This model allows to derive something one might identify as an "instant", to provide a sense in which these instants can be "ordered" and to introduce a convenient concept of "duration" measuring the separation between instants. Moreover, in this model, an intrinsic directionality — an evolution from past instants towards future instants – and thus an arrow of time is derived directly and automatically and does not have to be externally imposed. In particular, in Caticha's approach the arrow of time corresponds to an ordered succession of instants characterized by an increasing entropy. This set of concepts developed by Caticha constitutes what is called "entropic time", which is modelled as an ordered sequence of instants with the natural measure of duration chosen to simplify the description of motion. In his paper, Caticha shows that whether the entropic order, the inferred sequence of states of the particle-clock composite agrees with an objective order in an external "physical" time turns out to be quite irrelevant in the sense it is an empirically inaccessible question. The notion of a "physical time" is not needed. It is only the correlations among the particles and the clock that are observable and not their "absolute" order.

On the other hand, it is important to underline that a growing number of contemporary researchers are challenging the view that space-time is the fundamental arena of the universe. They point out that it does not correspond to a physical reality, and propose a "timeless space" as the fundamental arena instead. For example the English physicist Palmer, in his recent paper entitled *A New Geometric Framework for the Foundations of Quantum Theory and the Role Played by Gravity*, suggests that, since quantum theory is inherently blind to the existence of state-space geometries, attempts to formulate unified theories of physics within a conventional quantum-theoretic framework are misguided, and that a successful quantum theory of gravity should unify the causal non-Euclidean geometry of space-time with the timeless fractal geometry of state space [17]. Moreover, the Italian physicists Girelli, Liberati and Sindoni have recently developed a toy model in which they have shown how the Lorentzian signature (characteristic of special relativity) and Nordstroem gravity[2] (a diffeomorphisms

mechanics known today as Hamilton-Jacobi theory.

[2] In theoretical physics, Nordström's theory of gravitation was a predecessor of general relativity. Strictly speaking, there were actually two distinct theories proposed by the Finnish theoretical physicist Gunnar Nordström, in 1912 and 1913 respectively. The first was quickly dismissed, but the second became the first known example of a metric theory of gravitation, in which the effects of gravitation are treated entirely in terms of the geometry of a curved space-time. Neither of

invariant scalar gravity theory[3]) can emerge from a non-dynamical Euclidean space, with no diffeomorphisms invariance built in [18]. This toy-model provides an example where time (from the geometric perspective) is not fundamental, but simply an emerging feature. In more detail, this model suggests that at the basis of the arena of the universe there is some type of "condensation", so that the condensate is described by a four-dimensional manifold characterized by an Euclidean metric. Both the condensate and the fundamental theory are timeless. The toy model developed by Girelli, Liberati and Sindoni shows that at a fundamental level space is a timeless condensate, that time as humans perceive it is only an emerging feature and that different solutions of the equations of motion of the fields characterizing this condensate determine different metrics of the space-time background. If in a timeless background different metrics are possible and time represents only an emerging feature, this means just that, at a fundamental level, time cannot be considered a primary physical reality, and thus that the duration of material change has no existence of its own.

On the other hand, another Italian physicist, Enrico Prati of CNR of Agrate Brianza, has recently shown that Hamiltonian mechanics[4], both in the classical domain and in the quantum domain, is rigorously well defined without the concept of an absolute, idealized time and that in this timeless framework a physical system S, if complex enough, can be separated in a subsystem S_2 whose dynamics is described, and another cyclic subsystem S_1 which behaves as a clock. An important result of Prati's research is that, as a consequence of the gauge invariance[5], a complex system can be separated in many ways in a part

Nordström's theories are in agreement with observation and experiment. Nonetheless, the first remains of interest insofar as it led to the second. The second remains of interest both as an important milestone on the road to the current theory of gravitation, general relativity, and as a simple example of a self-consistent relativistic theory of gravitation. As an example, this theory is particularly useful in the context of pedagogical discussions of how to derive and test the predictions of a metric theory of gravitation.

3 In theoretical physics, the diffeomorphism invariance (known better as general covariance or general invariance) is the invariance of the form of physical laws under arbitrary differentiable coordinate transformations. The essential idea is that coordinates do not exist a priori in nature, but are only artifices used in describing nature, and hence should play no role in the formulation of fundamental physical laws.

4 Hamiltonian mechanics is one of the many equivalent ways to formulate mechanics.

5 The gauge invariance principle is one of the most important principles of invariance in physics. From a general point of view, it establishes that the description of a force is not altered by any modification of the length scales of rulers and of temporal scales of clocks utilized as measurement instruments. It is important to underline also that each fundamental interaction satisfies its peculiar gauge symmetry. For example, in the case of the electromagnetic interaction, the quantum-mechanical description of experiments on charged particles is invariant under local phase

which constitutes the clock and the rest [19]. This implies that the duration of material change provided by a cyclic subsystem cannot be considered a primary physical reality, that each subsystem which acts as clock measures only the numerical order of the dynamics of the other subsystem. According to the authors' point of view, Prati's research is a significant proof that indeed we live in a timeless background space and that clocks provide only the numerical order of material change inside this arena.

In virtue of the argumentations and the recent research that we have mentioned here it seems permissible to postulate that, at a fundamental level, universe is a timeless phenomenon: universe does not run in time, time is not part of the universe, clock/time is merely a measuring system for the numerical order of physical events. Duration of physical events has no existence on its own. Space-time is merely a mathematical model and is not the fundamental arena of the universe.

On the basis of the elementary perception and of the important results obtained by the eminent authors that we have suggested here, a timeless picture of the universe where time represents a "coordinate of motion" is thus legitimate to be considered. According to this view, the concept of the absolute, idealized time can be definitely abandoned and replaced with the numerical order $t_0, t_1, t_2 .. t_n$ of material change which is measured with a clock. The term "arrow of time" can be replaced with the term "arrow of numerical order". In this timeless picture, "before" and "after" exist only as a numerical order $t_0, t_1, t_2 ...... t_n$ of a physical event: the numerical order t_{n-1} is "before" the numerical order t_n just equivalently as natural number $n-1$ is "before" natural number n. The numerical order of material change is measured with "ticking" of a clock where t_0 represents beginning of the measurement, t_n end of the measurement. In all equations of physics the real meaning of symbol t is numerical order t_n of material change obtained with a clock.

transformations on the particle wave function, if one introduces a long-range field coupled to the charge – the electromagnetic field – and one makes simultaneously a suitable local gauge transformation on the electromagnetic potential.

1.2. Planck Time as a Fundamental Unit of Numerical Order. Physical Time and Mathematical Time

Let us consider a photon that is moving on the distance *d* from a point A to a point B of space. Let us suppose that the distance *d* between the point A and the point B is equal to 100 Planck distances. Then duration of movement of this photon from A to B is described by a numerical order given by 100 units of Planck time. Photon is not moving in time, it is moving exclusively in space. Time is numerical order of this movement. With clock one measures numerical order of motion of the photon from the point A to the point B.

If one assumes that space is granular at the Planck scale (this view is supported by some important theoretical results, in particular loop quantum gravity[6] as well as reticular space-time dynamics[7]), the smallest distance that photon can pass on the way from A to B is the Planck distance l_p. The fundamental unit to describe photon motion through a Planck distance l_p is Planck time $t_p = \sqrt{\frac{\hbar G}{c^5}}$ (where $\hbar$ is Planck's reduced constant, *G* is gravitation constant and *c* is the speed of light). Planck time exists only as a measuring system of numerical order of photon motion through a Planck distance. Photon position l_{p1} is "before" position l_{p2} in the sense that the numerical order t_{p1} corresponding to l_{p1} is "before" the numerical order t_{p2} corresponding to l_{p2}. Equivalently as natural number 1 is "before" natural number 2, t_{p1} is "before" t_{p2}. Velocity v of a material change is derived from its numerical order t_n: $v = \frac{d}{t_n}$ where *d* is the distance of the motion. Frequency γ of material change is derived from its numerical order t_n: $\gamma = \frac{1}{t_n}$.

---

[6] For a new interpretation of loop quantum gravity in the timeless space see chapter 2.11.

[7] Reticular space-time dynamics – developed by the Italian physicist Ignazio Licata since 1989 [20, 21] – is a model of space-time that fixes the most general features of the quantization of the background: it replaces the space-time continuum with a discrete structure of elementary space-temporal grains (each characterized by a non-linear oscillator) and implies that the motion of the particles, at a fundamental level, is not continuum but happens "at clicks".

One should distinguish between physical time (numerical order of material movements) and mathematical time (time as a parameter of physical theories). The time appearing in equations of physical theories is an abstract, mathematical parameter which is introduced in order to describe the physical time represented by the numerical order of material motions. The dynamics of matter, the material movements are the fundamental physical reality which are obtained from observation and measurement. The time appearing in the equations of Newtonian physics or of standard quantum mechanics can be never reached in experiments, does not ever appear in laboratory measurements. Only the numerical order of material motions, obtained with clocks, has a physical reality.

As photon moves in space only and not in time (intended as primary physical reality) in special relativity the fourth coordinate of Minkowski space-time x_4 is imaginary. It is a spatial coordinate given by the product of imaginary unit i, light speed c and numerical order t_n of an event: $x_4 = ict_n$. The fourth coordinate of special relativity is a mathematical coordinate which is introduced to describe the numerical order of a photon motion into space.

In special theory of relativity the fourth coordinate is therefore different from the first three coordinates. To our knowledge, this fact does not yet receive an adequate attention. The first three coordinates constitute physical realities because they are numbers which describe the position of a material object in space. The fourth coordinate is instead only a mathematical coordinate that it is used to describe the numerical order of events that run into three-dimensional physical space. As a consequence, the space-time manifold of special relativity can be considered as a "physical-mathematical reality".

It is more correct to consider the Minkowskian background of special relativity as a four-dimensional physical-mathematical reality where the fourth dimension is a mathematical coordinate which describes the numerical order of material change than a $3D+T$ space-time arena where both the three spatial coordinates and the temporal coordinate have a physical reality. Time exists only as a numerical order of material change that we measure with clocks. Fundamental unit of numerical order $t_0, t_1, t_2 ... t_n$ of material changes that run in space is Planck time $t_p = 5{,}39124 \cdot 10^{-44}$ seconds and is derived from the light speed: $t_p = \frac{c}{l_p}$ where l_p is a Planck distance. Planck time t_p exists in the universe as a fundamental physical unit that governs numerical order of material change.

Here time is given right definition according to our perception: material change runs in space only and not in time. Time is numerical order of material change and is measured with clocks. In the special relativistic equation $x_4 = ict_n$ the meaning of symbol t is numerical order t_n of material change obtained with a clock. In the Minkowskian formalism of special relativity, time t obtained with clocks is only a component of fourth dimension x_4 of a four-dimensional space having a physical-mathematical reality. At a fundamental level, material change runs in three-dimensional space. Universe is timeless in the sense that universe does not exist in an external time that flows on its own. Physical time is only a numerical order of change that runs in the universe. We can think in terms of "past" and "future" in the universe only in the sense of numerical order of material changes that run in a timeless three-dimensional space [22].

1.3. Special Relativity in a Three-Dimensional Euclid Space

In special relativity, the absolute, idealized time of Newton's physics is replaced by a class of related times: the Lorentz times of all the different Lorentz observers. In other words, the hypothesis of the existence of an idealized, absolute time is replaced by the hypothesis of the existence of the Minkowski manifold with its peculiar metric structure.

In the recent article *Special theory of relativity in a three-dimensional Euclid space* the authors of this book showed that Minkowskian mathematical four-dimensional space which satisfies Lorentz transformations (between the coordinates of an event measured by two observers, one at rest, the other in motion with constant speed parallel to the first spatial coordinate) can be replaced with a three-dimensional Euclid space with Galilean transformations for the three spatial dimensions and a law of transformation for the rate of clocks which implies that the speed of the moving clock (namely the proper time of the moving observer) does not depend on the spatial coordinates but is linked only with the speed of the inertial system associated with the moving observer. These equations are valid for both observers [23].

The view according to which the rate of moving clocks depends only on the speed of the moving observer turns out to be in agreement with some current research. For example, in some articles the Italian physicist Franco Selleri of Bari University introduced general transformations of space and time between inertial reference frames which seem to indicate clearly that space and time must be considered as separate entities: in special relativity time must be separated from space [24-26]. Given the inertial frames S_0 and S' endowed with Cartesian coordinates (where the origin of S', observed from S_0, is seen to move with velocity parallel to the first spatial axis of S_0), by starting from the following two empirically based assumptions:

1) the two-way velocity of light is the same in all directions and in all inertial systems;
2) clock retardation takes place with the usual velocity dependent factor when clocks move with respect to S_0,

Selleri found some general transformations for the space and time variables from S_0 to S' (called also equivalent transformations) which contain a free parameter e_1 which can be fixed by choosing a clock synchronization method in S'. Different values of e_1 are thus obtained from different synchronization conventions. The special theory of relativity (and thus the standard Lorentz transformations) is obtained for a particular choice of e_1. For all the values of e_1 except the case of the special theory of relativity, the system S_0 turns out to be a privileged reference frame. As it has been analyzed by Selleri, different values of e_1 determine different theories of space and time which are empirically equivalent to a large extent. But, as it was shown by Selleri, there are some particular phenomena (the accelerating spaceships, the rotating disk and the question of superluminal signals regarding the group velocity of electromagnetic radiation) that modify the situation to the point that the condition $e_1 = 0$ (corresponding to the absolute synchronization) becomes necessary. The adoption of $e_1 = 0$ leads to fundamental "inertial transformations" in which the transformation of the speed of clock does not contain the space variable. Selleri's theory of inertial transformations determines indeed an arena of special relativity in which the temporal coordinate must be clearly considered as a different entity with respect to the spatial coordinates just because the transformation of clocks' run between the two inertial systems does not depend on the spatial coordinates.

In our model of special relativity, time and space represent two distinct entities. In line with Selleri's results, the transformation of the speed of clocks proposed by the authors of this book determines an arena of special relativity in which the temporal coordinate must be considered as a different entity with respect to the spatial coordinates just because it does not depend on the spatial coordinates. Our results suggest that the three spatial coordinates of the inertial systems turn out to have a primary ontological status, define an arena that must be considered more fundamental than the standard space-time coordinates interpreted in the sense of Einstein. On the basis of the laws of transformation of the spatial coordinates and of the rate of clocks proposed by the authors, one can assume that the real arena of special relativity is not a mixed space-time stage but rather a three-dimensional space and that time does not represent a fourth coordinate of space but exists merely as a mathematical quantity measuring the numerical order of material changes. In the relativistic regime, clocks are measuring devices of the numerical order of material motions: each clock can be associated with a specific proper time of a specific observer. A clock as a measuring device of numerical order of material change in an experiment runs slower (generally, all material changes run slower) in a faster inertial system than in an inertial system at rest. Experiments with clocks in a fast airplane do confirm that these relative velocities are valid for the observers both of the faster inertial system and in the inertial system at rest.

Moreover, in the formalism proposed by the authors, there is no "time dilation" as is known in Einstein's standard special relativity. According to our approach, it is not true that dilation of time as a fourth coordinate of space causes clocks to have a slower rate. What really exists in different inertial systems is the relative velocity of material change (including run of clocks). In analogous way, there is neither a "length contraction" in the direction of motion of an inertial system along the first spatial axis. Length contraction along the first spatial axis would lead to a contradiction: in a moving inertial system horizontal photon clock positioned along the first spatial axis would shrink and would have faster rate than identical photon clocks positioned vertically. Regarding "length contraction" some other research leads to the same conclusions. Since 1905 when Einsteinian special relativity was published there has been no experimental data on "length contraction".

Moreover, in this model time travels into the past and into the future are out of question: one can travel in space only. Past, present and future exist only as a numerical order of changes which run in a three-dimensional space. In this perspective universe does not run in time, on the contrary time we measure with clocks is a numerical order of universal changes.

In synthesis, the interpretation of special theory of relativity suggested by the authors is based on the following postulates:

1) The velocity of light has the same value in all directions and in all inertial systems (postulate of the invariance of the velocity of light in the vacuum).
2) The fundamental arena of physical processes is a three-dimensional Euclid space where time exists merely as a mathematical quantity measuring the numerical order of material changes (and which can be defined as the proper time of the observer under consideration). Given two inertial frames o and o', where the origin of o', observed from o, is seen to move with velocity v parallel to the first spatial axis of the inertial frame o, the transformations between the spatial coordinates of these two systems are the standard Galilean transformations, while the transformation of the rate of clocks, namely between the proper times of the observes of these two inertial systems is described by a formalism which implies that the rate of moving clocks depends only on the speed of moving observers.

The interpretation of special relativity based on these two postulates introduce a new suggestive way to treat and analyse several relativistic phenomena, such as aberration of light, Jupiter's satellites occultation and radar ranging of the planets.

It is also interesting to make a comparison between this approach to special relativity and another approach, proposed recently by the American physicist Levy. The approach proposed by Levy in his article *Ether theory and the principle of relativity* implies the existence of a preferred ether frame, the variability of the one-way speed of light in the other frames and an alteration of the coordinates of reference frames as a consequence of the measurement distortions due to length contraction and clock retardation and of the synchronization procedures [27]. Instead, in the view proposed by the authors of this book, the transformations of the spatial coordinates and of the rate of clocks imply the following fundamental results: on one hand, there is no length contraction along the direction of motion of an inertial system and, on the other hand, time is a distinct entity from space and the idea of dilation of time can be replaced with the idea of relative velocity of material changes (including runs of clocks) in different inertial systems. The fundamental arena of special relativity is indeed a three-dimensional Euclid space subjected to Galilean transformations for the three coordinates of space and where time exists only as a mathematical parameter measuring the numerical order of material changes: in each inertial system there is a peculiar velocity of material changes. The view suggested by the authors presents Einstein's special relativity in an ordinary way without unnecessary introduction of the imaginary coordinate

$x_4 = ict$. In the universe time as a fourth dimension of space does not exist and so could not be "relative". With clocks we measure speed of material changes; relative is a velocity of material change running in a three-dimensional physical space.

1.4. Time as Numerical Order Resolves Zeno Problems on Motion

Zeno problems of motion confronted in terms of space and time are agitating human reason for centuries. As shown by the authors of this book in the paper [3], the concept of time as a numerical order of material change allows us to resolve Zeno's paradox of Achilles and the Tortoise in a clear and simple way.

In this paradox, the faster Achilles gives the Tortoise a head start in the race. But although Achilles can run 10 times faster than the Tortoise, he can never surpass the Tortoise because, for every distance unit that Achilles runs, the Tortoise also runs 1/10 that distance. So whenever Achilles reaches a point where the Tortoise has been, the Tortoise has also moved slightly ahead. Although the conclusion that Achilles can never surpass the Tortoise is obviously false, there are many different proposed explanations for why the argument is flawed.

If clocks provide only the numerical order of material motion, the paradox of Achilles and the Tortoise can be resolved by redefining velocity, so that the velocity of both runners is derived from the numerical order of their motion, rather than their displacement and direction in time. From this perspective, we can say that Achilles and the Tortoise move through space only, that Achilles surpasses Tortoise in space and not in a absolute time (that flows on its own). Velocity of both runners is derived from the numerical order of their motion. Let us imagine Achilles at the point A, Tortoise at the point T. Between A and T there is a distance d. When they start running into the same direction we activate a stopwatch. When Achilles is surpassing Tortoise we stop stopwatch. On the stopwatch we see $t_n = 10\,\text{sec}$. Achilles has passed 10 meters, his speed is $v = 1ms^{-1}$. Tortoise has passed 1 meter, its velocity is $v = 0{,}1ms^{-1}$. At the starting points the distance d between Achilles and Tortoise was 9 meters. Achilles runs distance $d_1 = 10$ meters. Tortoise runs distance $d_2 = 1$ meter. They both move in space only and not in time. Clock is a measuring device for the numerical order $t_0, t_1, t_2 ... t_n$ of their motion. Their velocities $v_a = \frac{d_1}{t_n}$ and $v_t = \frac{d_2}{t_n}$ are derived from the numerical order of their motion.

1.5. Time as Numerical Order Resolves Hypothetical Travel into Past and Twin Paradox

Since formulation of general theory of relativity scientific mind is challenged with idea of traveling into time. According to some interpretations of general relativity one could travel through a black hole into past. This allows preposition that during the "time travel" one could kill his grandmother. From that point of view he could not be born. This contradiction is resolved with understanding that one can travel in a three-dimensional timeless space only where physical time is numerical order of this motion. In this picture, traveling in a temporal dimension that flows on its own is out of question.

Also "twin paradox" charactering special theory of relativity is becoming clearer in the timeless approach where the fundamental arena is a three-dimensional space where time exists only as a numerical order of material change. Twin-brother in the fast spaceship is getting older slower than his brother on the planet earth. Velocity of material and biological change (and thus velocity of clocks) is slower in a fast space-ship than on the planet earth. In space-ship and on the planet earth material and biological change run in the same timeless three-dimensional space only and not in time. Twin and his brother are getting older in space only and not in time. Flow of time is flow of numerical order of their aging.

1.6. Psychological Time Creates Temporal Experience "Past-Present-Future"

Experiencing change of timeless universe through psychological time creates temporal experience in the linear process "past-present-future". Recent neurological research shows we experience flow of numerical order of material changes that run in a timeless space through psychological time "past-present-future". Temporal experience of material change "one after other in time" is result of neuronal activity of the brain which is basis of psychological time. In physical reality "one after other" exists only in a sense of numerical order of material changes occurring in a timeless space.

Through psychological time we experience the timeless nature of space as "present moment". Once we become aware of the psychological time we experience the timeless nature of space as "eternal present moment".

As regards the experience of material changes through psychological time "past-present-future", one can say that each moment is a complete, pre-existing and well defined entity: it is a specific version of the universe at that specific moment, with a specific version of the observer observing it at that moment. At each moment, we as observers have the experience of the present moment, and we have the memory of the past, and the expectations of the future. This is all part of one moment, one specific "time capsule" (if one wants to use Barbour's words). In the next moment the same thing is true: there is the experience of that moment, and the memory of the past, and the expectations of the future. From the physical point of view, all such moments, all such "time capsules" simply exist, timelessly. Such moments can be appropriately called 'now'. In this regard, Deutsch clearly underlines in his book *The fabric of reality*: "We do not experience time flowing, or passing. What we experience are differences between our present perceptions and our present memories of past perceptions. We interpret those differences, correctly, as evidence that the universe changes with time. We also interpret them, incorrectly, as evidence that our consciousness, or the present, or something, moves through time." [28]. From the physical point of view, universe is a timeless phenomenon whose only physical reality is the present moment, the "now". The past and the future of the universe have not an objective existence. Past and future of everything we see in the universe exist only in the sense of a numerical order of material changes. The only history of the evolution of the universe that has a sense from the physical point of view is a "history of now".

1.7. EPR Experiment and the Idea of Space as an Immediate Information Medium

The idea of time as a numerical order of material movements in a timeless space throws new light in the explanation and interpretation of Einstein-Podolski-Rosen (EPR) experiment and thus of non-local correlations between subatomic particles.

To be explicit and illustrate EPR experiment, let us consider an example given by the great American physicist David Bohm in 1951 [29]. We have a physical system given by a molecule of total spin 0 composed by two spin ½ atoms A and B in a singlet state (namely a state given by a quantum superposition, with equal weights, of the two following terms macroscopically distinguishable: one that indicates that the atom A has spin up and the atom B has spin down, the other that indicates that the atom A has spin down and the atom B has spin up). Given such a state, let us suppose to perform a spin measurement on the atom A in the z-direction and that we obtain the result spin up. Then, on the basis of the results of the usual quantum theory, the final wave function of the molecule indicates that in such situation the atom B has spin down. This means that by performing a spin measurement on the atom A, we know immediately the spin state of the atom B: if the measurement indicates that the atom A has spin up, the atom B turns out to have instantaneously spin down. But this outcome depends on the kind of measurement carried out on the particle A. By performing different types of measurement on the atom A we will bring about distinct states of the atom B, and this means that as regards spin measurements there are correlations between the two atoms. Although the two partial systems (the atom A and the atom B) are clearly separated in space (in the conventional sense that the outcomes of position measurements on the two systems are widely separated), indeed they cannot be considered physically separated because the state of the atom B is instantaneously influenced by the kind of measurements made on the atom A. Bohm's example shows clearly that entanglement in spin space implies non-locality and non-separability in Euclidean three-dimensional space: this comes about because the spin measurements couple the spin and space variables.

According to the view proposed by the authors of this book, the communication between subatomic particles in EPR-type experiments is immediate and the three-dimensional space (where time is not a primary physical reality but exists only as a numerical order of material changes) is the entity which determines this communication [30, 31]. The fact that, in Bohm's example, the state of the atom B is instantaneously influenced by the kind of measurements regarding the atom A suggests that quantum non-locality is determined by space, that in EPR experiment space is the information medium between elementary particles. There is no information signal in form of photon or some other particle traveling between atom A and atom B. The time of information transfer between atom A and atom B is zero. Information between atom A and atom B has not speed, has not duration: space itself is informing atom B about the behaviour of atom A and opposite. It is the three-dimensional space the medium which allows us to explain why and in what sense, in an EPR experiment, two particles coming from the same source and which go away, remain joined by a mysterious link, why and in what sense if we intervene on one of the two atoms A and B, also the other feels the effects instantaneously despite the relevant distances separating it. It is the three-dimensional space the fundamental arena which determines non-locality, which produces an instantaneous connection between the two particles as regards the spin measurements: by disturbing atom A, atom B may indeed be instantaneously influenced despite the big distance separating the two systems thanks to the three-dimensional space which puts them in an immediate communication.

In other words, one can say that in EPR-type experiments the three-dimensional physical space is a direct, an immediate information medium between elementary particles. In EPR-type experiments, in which we have two subatomic particles firstly joined and then separated and carried away one from the other, the behaviour of a subatomic particle is influenced instantaneously by the other particle thanks to space which functions as an immediate information medium.

1.8. Bohm's Quantum Potential and Its Link with the Idea of Space as an Immediate Information Medium

The idea of space as an immediate information medium between subatomic particles must not be considered completely new inside physics. One can find the important role of space as the medium determining quantum non-locality already in the context of Bohm's original quantum potential approach.

In his classic works appeared on "Physical Review" in 1952 and 1953, David Bohm introduced an alternative interpretation of quantum mechanics, known also as pilot wave theory, which can provide a causal description of atomic and subatomic processes [32, 33]. Bohm showed that if we interpret each individual physical system as composed by a corpuscle and a wave guiding it[8], the movement of the corpuscle under the guide of the wave happens in agreement with a law of motion which has the form of Newton's second law of motion of classical mechanics, with the only difference that here the particle is subjected, besides to a classical force, also to a quantum force, linked to an energy form called quantum potential[9]. In synthesis, the movement of an elementary particle, according to Bohm's pilot wave theory, is tied to a total force which is given by

---

[8] The idea that each subatomic particle can be interpreted as a physical system constituted by a corpuscle and a wave that guides it in the regions where the wave function is more intense (known also as wave-particle objective dualism) was originally proposed by the French physicist Louis de Broglie at the Solvay Conference in 1927.

[9] This denomination derives from the fact that in the mathematical expression of quantum potential Planck constant (the fundamental constant that characterizes the microscopic world) is present and there is a dependence on the wave function.

the sum of two terms: a classical force (derived from a classical potential) and a quantum force (derived just from the quantum potential). The main characteristic of Bohm's theory, which allows us to provide a causal description of atomic processes consists in the fact that here the wave function acts just as a pilot wave which guides the corresponding particle, through the action of the quantum potential, in the regions where it is more intense.

The quantum potential is the fundamental element which allows us to understand the features of Bohm's interpretation of quantum theory. If we examine the mathematical form of the quantum potential, we may note that it does not have the usual properties expected from a classical potential. In fact, the quantum potential depends on how the amplitude of the wave function varies in space and the action of this potential is like-space, namely creates onto the particle a non-local, instantaneous action. In a double-slit experiment, for example, if one of the two slits is closed the quantum potential changes, and this information arrives instantaneously to the particle, which behaves as a consequence. The fact that the action of quantum potential on a particle is like-space means that the three-dimensional space has clearly a relevant role in determining the motion of a subatomic particle. In virtue of Bohm's quantum potential, one can say that it is the three-dimensional space the medium responsible of the behaviour of quantum particles. Considering the double-slit experiment, the information that quantum potential transmits to the particle is instantaneous just because it is a spatial information. One can say that Bohm's quantum potential contains the idea of space as an immediate information medium in an implicit way.

It is also important to underline that in the standard interpretation of quantum mechanics the non-locality of quantum processes is an unexpected host and often does not receive the adequate attention. On the other hand, Bohm was the first to put in evidence in a clear way the origin of quantum non-locality. Bohm's theory manages to make manifest this essential feature of quantum mechanics, just by means of the quantum potential. In particular, taking into consideration a many-body system, Bohm's theory shows clearly that the quantum potential acting on each particle is a function of the positions of all the other particles and thus in general does not decrease with distance. As a consequence, the contribution to the total force acting on the i-th particle coming from the quantum potential does not necessarily fall off with distance and indeed the forces between two particles of a many-body system may become stronger, even if the modulus of the wave function may decrease in this limit. The equation of motion of the i-th in particle, in the limit of big separations states that the behaviour of a particle depends on the coordinates of all the n particles of the system: this determines just non-local correlations in a many-body system. In virtue of the features of quantum potential,

Bohm's theory turns out to be intrinsically holistic, in which "the whole is more than the sum of the parts". It is a merit of the pilot wave theory to show in such a direct way this property that, according to Bohm, "… is the newest and most fundamental ontological characteristic implied by quantum theory" [34].

The appearance of non-separability and non-locality in the Bohm approach led the great Irish physicist John Stewart Bell to develop a famous theorem [35], according to which a measurable quantity exists Δ which must satisfy the inequality $|\Delta| \leq 2$ whilst is predicted by quantum mechanics to have the value Δ =2,83[10]. Of course non-locality is not a feature that fits comfortably with the mechanical paradigm, but it was this feature that led Bohm to the conclusion that his approach was not mechanical.

Detailed investigations into these questions in the Bohm approach and in the review of other approaches to quantum mechanics led to the idea that the Cartesian order could no longer be used to explain quantum processes, in particular quantum non-locality and EPR-type experiments. What is needed is a radically new order in which to understand quantum phenomena.

In this regard, already in 1960 the American physicist Geoffrey Chew pointed out that there is no necessity to explain quantum processes on the basis of the space-time manifold [36]. This consideration of Chew appears legitimate if it is applied to the interpretation of EPR-type experiments. One encounters problems in explaining the instantaneous communication between subatomic particles if assumes that space-time is a fundamental entity. If space-time is assumed as a primary reality, then, ipso facto, locality should be absolute. Instead quantum particles show non-local correlations.

In 1980 Bohm suggested that the new order in which to understand quantum phenomena would be based on process and called this new order the implicate order [37]. According to this view, it is possible to recognize two levels of description in quantum mechanics: the standard interpretation and its formalism allows us to explore the explicate order of our measurements (or foreground); however, what happens in the explicate order represents a projection of a more fundamental level, namely the level of the implicate order (or background). The intention behind the introduction of the implicate order was simply to develop new physical theories together with the appropriate mathematical formalism that

[10] Bell's theorem can be considered the fundamental theoretical result which puts in evidence the non-local feature of quantum processes: the demonstration of Bell's theorem implies that an experience happened in the past between two subatomic particles creates between them a connection such that the behaviour of each of the two particles influences in a direct and instantaneous way the behaviour of the other particle independently on their distance.

will lead to new insights into the behaviour of matter and ultimately to new experimental tests. In this way Bohm in his last years departed from de Broglie's pilot wave: he suggested the necessity to consider non-locality as a primary fundamental characteristic of space-time, to introduce an intrinsic non-locality of the quantum world. The idea of the implicate order can be collocated just in this context. Bohm's work was practically directed towards an overcoming of the traditional role of space-time (which instead was present in de Broglie's original view), and to develop a theory of space-time where the quantum concepts appear as structural elements of the world which can be expressed through opportune topological constructions. In Bohm's view the non-locality is a characteristic subtended of space-time and the particles are seen as vibration modes of the global field which is the dynamical expression of the fundamental level, of the deep geometrical structure. Bohm's project was to develop a top-down approach: to introduce a global ontological structure and to try to obtain the form of the objects which emerge from this as manifestations of the undivided totality [38]. As regards his research on the implicate order, conducted mainly with the English physicist Basil Hiley, Bohm used very refined mathematical instruments, in particular directed his attention towards the non-commutative geometries, the non-linearity and the discreteness of the background of the universe.

Following this research line, Hiley suggested that quantum processes evolve not in space-time but in a more general space called pre-space, which is not subjected to the Cartesian division between res extensa and res cogitans. In this view, the space-time of the classical world would be some statistical approximation and not all quantum processes can be projected into this space without producing the familiar paradoxes, including non-separability and non-locality [39].

The considerations of Chew, the research of Bohm and Hiley clearly show the legitimacy to explain non-locality on the basis of approaches different from the conventional space-time manifold. Our view of space as an immediate information medium (and in which time exists only as a numerical order of material motion) can be considered the natural interpretative development of Bohm's and Hiley's research. According to the authors of this book, bohmian implicate order (or analogously Hiley's pre-space) can be assimilated to the idea of the three-dimensional physical space as an immediate information medium in the form of the special state of quantum potential. Since the quantum potential can explain the origin of non-locality and since according to the view here suggested the instantaneous communication between two quantum particles can be seen as an effect of the three-dimensional space which functions as a direct information medium, the possibility exists that there is a sort of correspondence between the quantum potential and the three-dimensional physical space, in particular that the quantum potential can be interpreted as the special "state" of the three-dimensional physical space in the presence of microscopic processes, and thus of quantum particles. When one takes into consideration an atomic or subatomic process (such as for example the case of an EPR-type experiment, of two subatomic particles, before joined and then separated and carried away at big distances one from the other), the three-dimensional physical space assumes the special "state" represented by quantum potential, and this allows an instantaneous communication between the two particles into consideration [30, 40].

In synthesis, one can say that in EPR-type experiments Bohm's quantum potential makes the three-dimensional physical space an "immediate information medium" between elementary particles. In EPR-type experiments the behaviour of a subatomic particle is influenced instantaneously by the other particle thanks to the three-dimensional space which functions as an immediate information medium in virtue of the geometric properties represented by the quantum potential.

But, what does it make indeed the three-dimensional space an immediate information medium in EPR-type correlations? If the space that we perceive seems to be characterized by local features, from which fundamental entity or structure does the property of the quantum potential to determine the action of the three-dimensional space as an immediate information medium derive? In this regard, according to the authors, it is important to mention that in the recent article *Bohmian split of the Schrödinger equation onto two equations describing evolution of real functions* the Russian physicist Sbitnev showed that the quantum potential can be interpreted as an information channel into the movement of the particles as a consequence of the fact that it determines two quantum correctors into the energy of the particle depending on a more fundamental physical quantity that can be appropriately called "quantum entropy" [41]. In Sbitnev's approach, the quantum entropy can be interpreted as the physical entity that, in the quantum domain, characterizes the degree of order and chaos of the vacuum – a storage of virtual trajectories supplying optimal ones for particle movement – which supports the density describing the space-temporal distribution of the ensemble of particles associated with the wave function under consideration [42]. This new way of reading bohmian mechanics can be called as the "entropic version" of bohmian mechanics or, more briefly, "entropic bohmian mechanics". In the entropic version of bohmian mechanics, the three-dimensional space distribution of the ensemble of particles describing the individual physical system under consideration generates a modification of the geometrical properties of the background space by determining two quantum correctors in the energy of the physical system under consideration (of the kinetic energy and of the potential energy respectively) without which the total energy of the system would not be conserved. With the introduction of the quantum entropy one can say that the action of the three-dimensional space as an immediate information medium derives just from these two quantum correctors to the energy of the system. As shown by the authors in the recent paper [43], by introducing the concept of the quantum entropy, the two quantum correctors to the energy of the system under consideration, depending on the quantum entropy, can be considered the ultimate elements which, at a fundamental level, produce an immediate information medium in the behaviour of the particles in EPR-type experiments. In this picture, the space we perceive turns out to be characterized by local features because in our macroscopic domain the quantum entropy has a peculiar behaviour (expressed by opportune mathematical conditions). Despite the view based on the quantum entropy is yet at first stages of development as regards the explanation of the local nature of the physical space in our everyday life, nonetheless according to the authors it can introduce and open interesting perspectives in the interpretation of the background of quantum processes. In particular, a crucial aspect of this entropic view lies in the fact that the quantum entropy, by producing two quantum corrector terms in the energy, can be indeed interpreted as a sort of intermediary entity between the three-dimensional space and the behaviour of quantum particles, and thus between the non-local action of the quantum potential and the behaviour of quantum particles. In synthesis, the ultimate source, the ultimate visiting card which determines the action of the three-dimensional space as an immediate information medium between quantum particles (and thus the non-local action of the quantum potential) is a fundamental vacuum defined by a quantum entropy which produces two opportune quantum corrector terms in the energy.

1.9. The Problem of Time-Symmetry. The Immediate Symmetric Interpretation of Quantum Non-Locality

Contrarily to what happens in Newtonian physics, the laws of standard quantum mechanics imply that by going backwards in time there is no guarantee that we obtain something that corresponds to what physically happens. If we imagine to film a quantum process backwards, namely inverting the sign of time, we are not sure that the reversal process will have the same path and numerical order. The standard quantum laws are not time-symmetric: by inverting the sign of the time, the filming of a given process could not have a description which agrees with the forward-time description.

The asymmetry of the standard interpretation is somewhat evident in the Schrödinger equation itself because the substitution $t \rightarrow -t$ yields a different equation. A more dramatic asymmetry concerns the collapse postulate; upon measurement, a wave function collapses to a state that implies a well-defined position of the particle under consideration only in the forward-time direction. The time-reverse of this process is not permitted. By inverting the sign of the time, the filming of a given process of collapse of a wave function does not allow us to reproduce what physically happens. The standard interpretation of quantum mechanics predicts therefore a dramatic disagreement between forward-time and reverse-time description of the same physical event. This fact would seem to provide an important motivation to search for an interpretation of quantum mechanics in which a forward-time and reversed-time perspective of the same physical events would be interpreted in the same manner.

Also Bohm's original quantum potential approach, although allows us to obtain a causal description of atomic processes and to provide a satisfactory explanation of quantum non-locality in a coherent and compatible way with the idea of space as an immediate information medium, like the original standard quantum mechanics, meets problems by inverting the sign of the time. Also in the context of Bohm's theory, by inverting the sign of time, there is no guarantee that we obtain something that corresponds to what physically happens. In fact, the quantum potential, although has a like-space, an instantaneous action, however comes from Schrödinger equation which is not time-symmetric and therefore its expression does not seem completely satisfactory – from the mathematical point of view – if one imagines to film a quantum process backwards. Thus, the original bohmian approach, although allows us to explain in a clear way quantum non-locality, cannot be considered completely convincing, under the point of view of the symmetry in time, because is not time-symmetric.

It is however important to underline that in the picture of the universe suggested in this book, as time exists only as a numerical order of material change occurring in a timeless space, the problem of the symmetry in time of physical laws and theories does not seem of relevant importance from the physical point of view. As experimental data confirm that time exists only as a numerical order of material change occurring in a timeless space, that time as flow of numerical order of material change is therefore irreversible, the question regarding the symmetry in time of physical laws turns out to be secondary on the physical point of view. It has not primary physical relevance to see what happens to our theories by inverting the sign of time because it is not possible to film a physical process backwards.

In the approach suggested in this book, according to which time is a measuring system of the numerical order of material changes and movements occurring in a timeless space, the question of the symmetry in time of physical laws must be considered a purely mathematical matter. Clearly, mathematical models can be continuously improved, and under this point of view the problem of the symmetry in time of physical theories has sense to be taken into consideration, but only from the point of view of the mathematical formalism, as an attempt to find a more general and unitary mathematical scheme. From the physical point of view, in virtue of the fact that time exists only as a numerical order of material change and the stream of numerical order of material change is therefore irreversible, the problem of the symmetry in time indeed has not a primary significance. On the basis of these considerations, we can say that the question to interpret in the correct way, in symmetric terms in the exchange of t in $-t$, every quantum process and to develop thus a mathematical extension of standard quantum mechanics (or of Bohm's quantum potential approach) that can be symmetryzed by inverting the sign of the time, must be faced only from the mathematical point of view.

Since 2008 the authors of this book have really suggested a mathematical, symmetryzed extension of Bohm's quantum potential which can explain both the forward time process and the time reverse process of quantum processes [30]: this symmetryzed extension of Bohm's quantum potential can be considered a better candidate than Bohm's original quantum potential in the interpretation and explanation of quantum phenomena (and also of quantum non-locality) from the mathematical point of view. Instead, from the physical point of view, as time as flow of numerical order of material change in a timeless space is irreversible, this symmetrized extension of Bohm's approach cannot be considered as a primary reality and the original Bohm's approach acts well. In other words, we can say that, in order to reproduce and explain quantum phenomena (including quantum non-locality), the symmetrized quantum potential is a mathematical reality, while the original Bohm's quantum potential is a physical reality.

In this chapter the aim is to synthesize the most important results regarding our research about the symmetryzed version of quantum potential, able thus to explain, from the mathematical point of view, a symmetric and instantaneous communication between subatomic particles.

In this regard, before all, it is important to underline that the mathematical problem of time-symmetry in standard quantum mechanics can be resolved by taking into consideration the time-symmetric formulation of quantum mechanics developed by the American physicist of the San José State University K.B. Wharton in 2007 [44]. In Wharton's model, the wave function is no longer a solution to the standard Schrödinger equation, but instead is the solution containing two components (the one for the forward-time process and the other for the time-reverse process) to a time-symmetric extension of Schrödinger equation. Wharton's model is therefore based on two physical equations, the one for the forward-time processes (which is the standard Schrödinger equation), the other for the time-reversed processes (which is the time-reversed Schrödinger equation). Moreover, as regards the analysis of the measurement processes, instead of a collapse postulate, Wharton's formulation imposes a boundary condition on the wave function at every measurement, equal to the outcome of that measurement. In virtue of its features, Wharton's model does not require a time-asymmetric collapse of the wave function upon measurement. Therefore it can be considered an interesting starting-point in order to interpret in the correct manner, from the mathematical point of view, both the forward-time and the reversed-time perspectives of the same physical event inside standard quantum mechanics.

Now, since non-locality is due to bohmian quantum potential, to the like-space, instantaneous action of the quantum potential, in order to assure the symmetry in time needed to interpret, from the mathematical point of view, also the time-reverse process in the correct manner and thus to find the most appropriate and complete mathematical candidate for the state of space as a direct information medium between subatomic particles, we have reformulated Bohm's approach for the time-symmetric equation introduced by Wharton. In this way we have obtained a symmetryzed extension of Bohm's version of quantum mechanics which is characterized by a symmetryzed quantum potential at two components (the one depending on the wave function characterizing the forward-time process – and which coincides with the original Bohm's quantum potential – and the other depending on the wave function characterizing the reversed-time process). The symmetryzed quantum potential can be considered the starting-point to have a symmetry in time in bohmian quantum mechanics [30, 45].

Just like the quantum potential of the original Bohm theory, also the symmetryzed quantum potential that we have introduced has an action which is stronger when the mass is more comparable with Planck constant, and which is like-space, non-local, instantaneous. The difference with respect to the original bohmian mechanics lies in the fact that the symmetrized quantum potential has two components and therefore its space-like, non-local, instantaneous action is predicted not only by the forward-time process but also by the time-reverse process. In this picture, the original quantum potential can be seen as the first component of the symmetryzed quantum potential: it is the component that explains the forward-time quantum processes (and thus, as time as flow of numerical order of material change is irreversible, must be intended as the real physical component). The second component describes the time-reverse process and thus must be intended only as a mathematical component, that has not a primary physical reality because it is not possible to go backwards in the physical time intended as numerical order of events.

On the other hand, we found that the second component – regarding the time-reverse processes – of the symmetrized quantum potential turns out to have an opposed sign with respect to the first component: this is just a consequence of the fact that the measurable time exists only as a measuring system of the numerical order of material change. In other words, in the symmetrized quantum potential approach, the mathematical features of the second component of the symmetrized quantum potential imply just that it is not possible to go backwards in the physical time intended as numerical order of physical events.

According to the symmetrized quantum potential approach, in an EPR-type experiment, the three-dimensional space assumes the special "state" represented by the first component of the symmetrized quantum potential, and it is this special state of space that, according to the authors, determines a non-local, instantaneous communication between the particles into consideration. Instead, the second component of the symmetrized quantum potential regards the reversed-time processes, imply that time as flow of numerical order is irreversible, and thus must be considered only a mathematical entity: it allows us to explain quantum phenomena, in particular non local correlations in many-body systems – and thus EPR experiments – from the mathematical point of view in the most correct, coherent and complete way (namely from the point of view of the mathematical symmetry in time). The symmetrized quantum potential at two components (the one for the forward-time process, the other for the time-reverse process) can be considered the most general appropriate candidate to provide a primary mathematical reality to a three-dimensional space intended as a direct information medium. While the standard Schrödinger equation and the original bohmian theory imply that by filming back a physical event it is not possible to see what effectively happens (namely the real interaction which characterizes the system under consideration), with the introduction of the symmetrized quantum potential we can interpret in the correct way, from the mathematical point of view, both the forward-time process and the time-reverse process and thus the numerical order of a quantum process and the idea of the three-dimensional space as a direct information medium can receive a primary mathematical significance. It is by means of the special state represented by the symmetrized quantum potential, that in the subatomic world at a fundamental level space can be interpreted as an immediate information medium, as an entity that puts the particles into an immediate contact, from a mathematical point of view. More precisely, we can say that, in the presence of quantum particles, space assumes the physical state represented by the first component of the symmetrized quantum potential (that coincides with the original Bohm's quantum potential), and it is this special state of space that, according to the authors, determines a non-local, instantaneous communication between the particles into consideration. As it is not possible to go backwards in the physical time intended as numerical order of material change, it is first component of the symmetrized quantum potential (which coincides with the original Bohm's quantum potential) that must be considered the most appropriate candidate to provide a physical reality to space as a direct information medium. In EPR experiment it is just the first component of the symmetrized quantum potential (namely the original Bohm's quantum potential) that makes physical the three-dimensional space an "immediate information medium" which keeps two elementary particles in an immediate contact. Instead the second component is only a mathematical component: it reproduces what physically happens if one would imagine to film a quantum process backwards and is the mathematical translation of the fact that measurable time exists only as an irreversible numerical order of events. It allows us to explain quantum phenomena, in particular non local correlations in many-body systems – and thus EPR experiments – in the most correct and complete way from the mathematical point of view (namely from the point of view of the mathematical symmetry in time).

Moreover, both the components of the symmetrized quantum potential are physical quantities which derive from an opportune quantum entropy. More precisely, the first component of the symmetrized quantum potential corresponds to a quantum entropy defining the degree of order and chaos of the vacuum for the forward-time processes; the second component of the symmetrized quantum potential corresponds to a quantum entropy defining the degree of order and chaos of the vacuum for the time-reverse processes. The non local quantum geometry characterizing the instantaneous communication between subatomic particles in EPR-type experiments can be thus defined by a symmetrized quantum entropy characterized by two components which determine the two components of the symmetrized quantum potential. In EPR-type experiments the action of the three-dimensional space as an immediate information medium derives just from the two quantum correctors to the kinetic energy and the potential energy of the system under consideration, both for the forward-time processes and for the reversed-time processes, which are associated with the symmetrized quantum entropy. The symmetrized quantum entropy can be therefore considered the fundamental entity which, at a fundamental level, produces an immediate information medium in the behaviour of subatomic particles in EPR-type experiments. In synthesis, in EPR-type experiments (and more in general in immediate physical phenomena regarding the quantum domain) a three-dimensional timeless space acts as an immediate information medium in the sense that, as a consequence of the symmetrized quantum entropy, the first component of the symmetrized quantum potential makes physical space an "immediate information medium" which keeps two elementary particles in an immediate contact (while the second component of the symmetrized quantum potential reproduces, from the mathematical point of view, the symmetry in time of this communication and the fact that time exists only as a numerical order of material change). We can call this peculiar interpretation of quantum non-locality (based on the idea of space as an immediate information medium) as the "immediate symmetric interpretation" of quantum non-locality [45].

1.10. Immediate and Non-Immediate Physical Phenomena

The "immediate symmetric interpretation" of quantum non-locality determines important consequences as regards the arena of the universe. If there are quantum events which can be explained through the idea of the direct, immediate information transfer, and if, in virtue of the peculiar features of the first component of the symmetrized quantum potential (namely of the original Bohm's quantum potential), at the fundamental level of quantum processes space functions as an immediate information medium, this means, from the physical point of view, that the background timeless space of quantum processes can be considered as an absolute reference frame. In other words, the timeless quantum space which acts as a direct information medium (and which is associated with the first component of the symmetrized quantum potential, namely with Bohm's quantum potential) defines a absolute reference system in order to study the quantum phenomena. All physical events must be analyzed with respect to this absolute reference frame represented by the timeless quantum space which acts as an immediate information medium.

The view of time as a coordinate of motion, as a numerical order of material changes occurring in timeless background space introduces therefore interesting perspectives as regards the ultimate arena and the ultimate nature of physical events. In the approach based on a timeless space acting as an immediate information medium, physical phenomena can be divided into two basic groups: absolute phenomena (phenomena that are characterized by a constant velocity) and relative phenomena (phenomena that are characterized by a velocity which is relative on the inertial system and on gravity) [46]. The absolute physical phenomena can be divided in the following two categories:

- absolute physical phenomena characterized by a numerical order equal to zero, that are function of timeless space itself (called also immediate physical phenomena);
- absolute physical phenomena of electromagnetic nature that have a light speed and are characterized by a numerical order $t_n > 0$.

The relative phenomena are instead physical phenomena characterized by a numerical order $t_n > 0$ that are function of particles' motion (particles that are bigger than photon, and thus characterized by a speed less than light speed) or bodies' motion into timeless space.

Quantum events which happen in no time, namely which are characterized by a numerical order equal to zero, and can be explained through the idea of space as a direct information medium, can be therefore considered as absolute physical phenomena. These particular physical events are independent on the inertial system and on gravitation: they depend only on the timeless space itself which acts as an immediate information medium and for them the two theories of relativity are not valid. The non-local correlations between quantum particles in EPR-type experiments and other immediate physical phenomena like tunnelling or the quantum communications regarding the continuous variable systems or the quantum excitations from one atom to another in Fermi's two-atom system (which we will be treated in major detail in chapter 1.11), can be considered as significant examples of absolute physical phenomena with a numerical order equal to zero that are functions of timeless space itself [47-49, 17]. Their velocity is always constant, namely zero and is absolute (just the light speed of special relativity). The speed zero in the quantum domain turns out to have a ontological status similar to the light speed of special relativity.

Moreover, the background timeless space which acts as a direct information medium determines another important consequence: the real physical speed of a physical event must be measured with respect to the reference frame associated with this timeless space. If the speed of an event turns out to be lower than light speed (with respect to this background space) this means that the event into consideration cannot be considered absolute, but is relative on the inertial system and on gravity and thus for this event relativity works. So, quantum processes which are characterized by a material change and thus by a numerical order $t_n > 0$ and by a speed lower than light speed (with respect to this timeless background) can be considered relative physical phenomena: they are physical phenomena whose speed depend on the inertial system and on gravity (and thus whose speed is a function of other physical systems chosen as reference frames).

1.11. Immediate Physical Phenomena and Their Explanation Through the Idea of Space as an Immediate Information Medium

EPR-type experiments and thus non-local correlations between subatomic particles can be considered the fundamental immediate physical phenomena occurring in the quantum domain. In the timeless approach developed by the authors of this book, in EPR-type experiments the elapsed time for quantum entanglement is zero and the instantaneous communication between the particles under consideration is determined by the three-dimensional space which acts as an immediate information medium. Bohm's original quantum potential (or, that is the same, the first component of the symmetrized quantum potential) can be considered the better candidate to represent, from the physical point of view, the concept of space as an immediate information medium and therefore to explain EPR-type experiments. The symmetrized quantum potential (characterized by two components) can be considered the better candidate to represent mathematically the concept of space as an immediate information medium (the first component can explain and reproduce the forward-time processes, while the second component can explain and reproduce the time-reversed processes).

It is interesting to underline that, in the quantum domain, there are however also other relevant immediate physical phenomena (and all these phenomena are somewhat linked with the fundamental immediate physical event of the quantum domain, namely quantum non-locality).

A relevant immediate physical event of the quantum domain is represented by tunneling processes. In the recent article *Attosecond Ionization and Tunneling Delay Time Measurements in Helium* by Eckle of the university of Zurich and other researchers, a conclusion is drawn that "an electron can tunnel through the potential barrier of a He atom in practically no time" [47]. It is well established that electrons can escape from atoms through tunneling under the influence of strong laser fields, but the timing of the process has been controversial and far too rapid to probe in detail. The authors of this article used attosecond angular streaking to place an upper limit of 34 attoseconds and an intensity-averaged upper limit of 12 attoseconds on the tunneling delay time in strong field ionization of a helium atom. The ionization field derives from 5.5-femtosecond-long near-infrared laser pulses with peak intensities ranging from 2.3×10^{14} to 3.5×10^{14} watts per square centimeter. The technique relies on establishing an absolute reference point in the laboratory frame by elliptical polarization of the laser pulse, from which field-induced momentum shifts of the emergent electron can be assigned to a temporal delay on the basis of the known oscillation of the field vector.

Moreover, as regards tunneling the Chinese physicist Whong Chao Wou affirms: "The physical meaning of the zero real time is that the particle instantly jumps from one side of the barrier to the other regardless of the thickness. This leads to the illusion that tunneling particles could actually travel faster than light. The results of recent experiments in quantum optics concerning tunneling time can be thought of as the first experimental confirmation of the existence of imaginary time. Relativity is not violated" [50].

The interpretation here proposed is that tunneling particles cannot travel in mathematical imaginary time *it*, they simply experience no time as they transfer from one measurable location to another. Information transfer of tunneling particle from one side of the barrier to the other is carried directly by the quantum space which acts as an immediate information medium (in the special physical state represented by the first component of the symmetrized quantum potential, namely Bohm's quantum potential).

On the other hand, not only tunneling phenomena seem to indicate that timeless quantum communication is a real phenomenon. For example, in the article of 2005 *The Capacity of a Quantum Channel for Simultaneous Transmission of Classical and Quantum Information*, the American physicists of the university of Southern California Devetak and Shor underline: "An expression is derived characterizing the set of admissible rate pairs for simultaneous transmission of classical and quantum information over a given quantum channel, generalizing both the classical and quantum capacities of the channel. Although our formula involves regularization, i.e., taking a limit over many copies of the channel, it reduces to a single-letter expression in the case of generalized dephasing channels. Analogous formulas are conjectured for the simultaneous public-private capacity of a quantum channel and for the simultaneously 1-way distillable common randomness and entanglement of a bipartite quantum state" [51]. In recent years, quantum communication protocols have been extended to the domain of continuous variable systems, i.e., quantum systems, like the bosonic modes of the radiation field, which are characterized by infinite dimensional Hilbert spaces. In particular, these protocols show how a sender (say Alice) can exploit bosonic modes in order to send similar signals to a receiver (say Bob) and then extract a secret binary key from these signals. Beyond the possibility of such a continuous variable quantum key distribution, it has been shown how to use these systems in order to perform a (quasi)confidential quantum direct communication, i.e., the (quasi)private communication of a message from Alice to Bob which is directly encoded in continuous variable systems. In particular, in the recent paper *Quantum direct communication with continuous variables*, the Italian physicist of MIT Pirandola and others write: "We show how continuous-variable systems can allow the direct communication of messages with an acceptable degree of privacy. This is possible by combining a suitable phase-space encoding of the plain message with real-time checks of the quantum communication channel. The resulting protocol works properly when a small amount of noise affects the quantum channel. If this noise is non-tolerable, the protocol stops leaving a limited amount of information to a potential eavesdropper" [48].

Another relevant immediate physical event is then represented by Fermi's two atoms system. Let A and B be two atoms or, more generally, a ''source'' and a ''detector'' separated by some distance *R*. At *t*=0 A is in an excited state, B is in its ground state, and no photons are present. As it has been shown by the German physicist Gerhard C. Hegerfeldt of the university of Göttingen, a theorem can be proved that in contrast to Einstein causality and finite signal velocity the excitation probability of B is nonzero immediately after *t*=0 [49].

It is interesting to observe that space as the "direct information medium" in the form of the symmetrized quantum potential allows us to resolve in clear way the causality problem of Fermi's two atoms system. In fact, according to the immediate symmetric interpretation of quantum non-locality, excitation probability of B is nonzero because the physical space in which atoms exist functions as a "direct medium of excitation". Excitation from atom A to atom B is direct and immediate via space through its special state represented by the first component of the symmetrized quantum potential, and not indirect via intermediate particles which move in space from atom A to atom B.

Moreover, as regards the causality problem of Fermi's two atom system, it is important to point out that standard quantum mechanics can reproduce mathematically only the forward-time process of the excitation from atom A to atom B. Instead, in the symmetrized extension of Bohm's theory, the special mathematical state represented by the symmetrized quantum potential allows us to reproduce in the correct way, from the mathematical point of view, also the time-reverse of the excitation from atom A to atom B, namely just through the idea of the direct information. The excitation from atom A to atom B arrives instantaneously thanks to the direct, immediate action of space in its special state represented by the first physical component of the symmetrized quantum potential, while the second component, by allowing us to explain in the correct way also the time-reverse of this process, namely filming back the process of excitation from atom A to atom B, provides a mathematical symmetry in time to the process and thus provides a time symmetric mathematical reality to the idea of the direct action of space.

In synthesis, here it is considered the perspective that in EPR-type experiments (and in all immediate physical events regarding the quantum domain) information does not move through space-time, but instead moves through a timeless space background (in its special state represented by the first component of the symmetrized quantum potential, namely Bohm's quantum potential), more precisely that space in which particles exist can be considered as a direct information medium between the entangled quanta. In other words, quantum non-locality (and all immediate physical events regarding the quantum domain) introduce the perspective that, at a fundamental level, space in which particles exist acts as a direct information medium (in its special state represented by the first component of the symmetrized quantum potential, namely Bohm's quantum potential).

If in the subatomic world the symmetrized quantum potential is the best candidate to express in a mathematical picture the idea of physical space as a real direct medium of information transfers between elementary particles, in a complete physical theory the possibility is opened that a fundamental arena in which space functions as a direct information medium is the primary element from which every field and every object of physics derives and which can reproduce the fundamental interactions and physical fields in a unified way. If non-locality is considered as the essential characteristic of the physical world and the idea of the symmetrized quantum potential at two components seems the most general and consistent way to introduce from the mathematical point of view non-locality (where the first component, namely Bohm's original quantum potential, is the entity which presents the physical idea of space as an immediate information medium), in a fundamental theory the symmetrized quantum potential should assume a crucial role as a mathematical reality of the processes (and in particular its first component should represent the ultimate physical reality) and all the objects of physics might emerge from it as special states. In particular, the possibility is opened that there is an important link between this fundamental arena and the Planck scale, in particular with the granular structure of space predicted by loop quantum gravity. We consider the perspective that at the Planck scale information and energy transfers are immediate. On the basis of the immediate interpretation (and of the mathematical immediate symmetric interpretation), the perspective is opened that direct quantum information transfers run over quanta of space which have the size of Planck's length. According to this interpretation, at the Planck scale space acts as an immediate information medium. At this scale information transfers are immediate: elapsed time for them to happen is therefore zero.

1.12. A Three-Dimensional Timeless Quantum Vacuum as a Fundamental Bridge Between Gravitation and the Quantum Behaviour

The laws of quantum mechanics applied to electromagnetic radiation imply the existence of a fundamental level, a background "sea" of electromagnetic zero-point energy that can be called the electromagnetic quantum vacuum. The introduction of a quantum vacuum as underlying structure of the quantum universe suggests that the processes regarding the observable forms of matter and energy in the three-dimensional space we perceive emerge from this more fundamental and deeper background. Several theoretical results – being based either on laboratory experiments or on astronomical observations – indicate that the vacuum energy density should be non-zero.

The notions of physical vacuum originated after the birth of quantum mechanics, in connection with the development of the idea of spontaneous emission of an isolated excited atom. In general, a quantum theory of fields associates to each point of space-time a creation/annihilation operator, in other words a linear oscillator whose activity, because of the uncertainty principle, is never zero. This leads to define a zero-point field, namely a fundamental energy of the quantum vacuum, from which virtual fluctuations arise.

The quantum vacuum predicted by the quantum theory of the electromagnetic field (like in other quantum field theory) is a real dynamic entity, able to manifest energetic fluctuations (described in terms of virtual particles), not directly observable but significant in the investigation of the behaviour of physical systems because it can produce real and observable effects on them. For example, it was found that the polarization of the electromagnetic component of the physical vacuum could manifest itself in the spatial "smearing" of the electron and a change, as a result, of the potential energy of its interaction with the nucleus, thus providing conditions for the removal of the degeneracy of the energies of the 2*S* ½ and 2*P* ½ states in the hydrogen atom. This phenomenon of the orbital shift, with consequent variation of energetic levels – known as Lamb shift – is just owed to the action of virtual particles, in other words of the fluctuations of the quantum vacuum [52]. It was also demonstrated that the quantum fluctuations of the electromagnetic component of the physical vacuum in regions contiguous with material objects could alter the relativistic quantum relationships in the near-surface regions of the objects and thus give rise to its macroscopic manifestations, namely the Casimir effect[11] and the Josephson contact noise[12]. All the effects here mentioned, being of electromagnetic nature, vanish if the fine structure constant $\alpha = \frac{e^2}{\hbar c}$ (where *e* is the electron charge, *c* is the velocity of light in vacuum and $\hbar$ is Planck's reduced constant) is negligible [53-55].

The development of quantum chromodynamics for the study of the elementary constituents of nuclear particles and their interactions brought in a fresh knowledge of the nature of the physical vacuum on high-energy scales. It turned out in particular that at an energy density of $E_{QED} \approx 200MeV$ a confinement-deconfinement transition was realized inside the nucleus (quarks were no longer bound in nucleons) to form a quark-gluon plasma or quark soup. The strong interaction constant α_S in that case proved to be dependent on the excitation energy: its magnitude changed from $\alpha_S \approx 1$ at low energies to $\alpha_S \approx 0{,}3$ at energies of a few gigaelectron-volts, depending but weakly on energy thereafter [56].

In the last decade, the notions of physical vacuum have come into wide use in cosmology [57-60]. It is precisely the physical vacuum that is associated, in the standard model of the dynamics of the universe, at whose root lie the Friedmann

---

[11] In quantum field theory, the Casimir effect, named after the Dutch physicist Hendrik Casimir, is a physical phenomenon arising from a quantized field. The typical example is of two uncharged metallic plates in a vacuum, placed a few micrometers apart. By studying this field in the picture of the electromagnetic component of the quantum vacuum, it is seen that the plates affect the virtual photons which constitute the field, and generate a net force, either an attraction or a repulsion depending on the specific arrangement of the two plates. Because the strength of the force falls off rapidly with distance, this phenomenon is measurable only when the distance between the objects is extremely small.

[12] The Josephson noise is the phenomenon of supercurrent — namely a current that flows indefinitely long without any voltage applied — across a device known as a Josephson junction, which consists of two superconductors coupled by a weak link.

equations of the general theory of relativity[13], with "dark energy" that accounts for 73% of the entire energy of the universe.

It is of high theoretical and practical interest to be able to calculate the actual value of the vacuum energy density. In this regard, according to the authors of this book, the Planckian metric (the system of units – constituted by Planck's length, Planck's time and Planck's mass – introduced by Max Planck in 1899 from dimensionality considerations by combining the three fundamental physical constants, namely the speed of light c, the Planck constant h, and the Newton's universal gravitational constant G) suggests an interesting way to define a quantum vacuum energy density. By starting with a semi-classical argument derived from orbital mechanics, and then by imposing on it the necessary constraints of relativity and of quantum mechanics, the authors found that, on the basis of Planck's metric, the maximum energy density characterizing the minimum quantized space constituted by Planck's volume is given by the Planck energy density $\rho_{pE} = \frac{m_p \cdot c^2}{l_p^3} = 4{,}641266 \cdot 10^{113} \frac{Kg}{ms^2}$ (where m_p and l_p are Planck's mass and Planck's length respectively) and this defines a universal property of space [61].

On the other hand, Italian physicists Leonardo Chiatti and Ignazio Licata developed an interesting approach, known as model of transactions, which implies that the fundamental arena of the universe is an archaic, atemporal vacuum that constrains and conveys the dynamical processes we observe [62-63]. In this picture, the vacuum is seen as a fabric from which patterns emerge by reduction-state processes and such patterns influence the vacuum activity, in a quantum feedback. According to Chiatti's and Licata's approach the only truly existent "things" in the physical world are the events of transactions, namely the events of creation and destruction (or, in other words, physical manifestation and demanifestation) of certain qualities. Such properties' measurement is all that we know of the physical world from an operational point of view. In this picture, any other construction in physics – like the continuous space-time notion itself or the evolution operators – has the role to causally connect the measured properties: they are "emergent" with respect to the network of events. The connection between the events of creation or destruction is possible because these events derive from the transformation of the same atemporal vacuum. In this picture, it looks as if every quantum process and time itself emerges from this invariant atemporal substratum and is re-absorbed within it.

---

13 The Friedmann equations are a set of equations in cosmology that describe the expansion of space in homogeneous and isotropic models of the universe within the context of general relativity.

By starting from the Planckian metric, the authors of this book recently developed a model of a three-dimensional isotropic quantum vacuum composed by elementary packets of energy having the size of Planck's volume and whose most universal physical properties are the energy density and state-reduction processes (of creation-annihilation of quantum particles) analogous to Chiatti's and Licata's transactions [64, 65].

In the free space, in the absence of matter, the energy density of the three-dimensional quantum vacuum is at its maximum and is given by the Planck energy density. In the ordinary space-time we perceive, the appearance of matter corresponds to a given change of the energy density of the three-dimensional quantum vacuum. The presence of a given massive particle or massive object in our level of physical reality derives from a diminishing of the energy density of quantum vacuum.

In analogy with Chiatti's and Licata's approach, in the three-dimensional quantum vacuum model suggested by the authors of this book the events of preparation of the initial state (creation of a particle or object from the three-dimensional quantum vacuum) and of detection of the final state (annihilation or destruction of a particle or object from the three-dimensional quantum vacuum) are assumed as the two only real primary physical events. The connection between these two events is ensured by their common origin in the timeless background represented by the three-dimensional quantum vacuum. In this picture, the process amplitudes are mathematical devices suited to calculate the statistics of the connection itself. Accordingly, the virtual processes obtained by expanding these amplitudes into partial amplitudes are also mathematical devices. Moreover, the evolution of a massive particle – which is originated from the three-dimensional quantum vacuum – is determined by appropriate waves of the vacuum associated with the amplitudes of creation or destruction events. The waves of the vacuum act in a non-local way through an appropriate quantum potential of the vacuum (which, so to speak, guides the occurring of the processes of creation or annihilation in the three-dimensional quantum vacuum). Just like the amplitudes of creation and destruction events, also the quantum potential of the vacuum must be considered as a mathematical entity. Indeed, the quantum potential of the vacuum can be seen as the primary mathematical reality which emerges from the very real extreme primary physical realities, namely from the processes of creation and annihilation. In virtue of the primary physical reality of the processes of creation and annihilation and of the non-local action of the quantum potential which is associated with the amplitudes of them, in the three-dimensional quantum vacuum the duration of the processes from the creation of a particle or object till its annihilation has not a primary physical reality but exists only in the sense of numerical order. In other words, in the three-dimensional quantum vacuum time exists merely as a mathematical parameter measuring the dynamics of a particle or object.

The two only primary extreme physical events of the three-dimensional quantum vacuum (namely the creation and annihilation of an elementary particle) are each corresponding to a peculiar reduction of a state vector. The authors of this book have called them "RS processes" where RS stands for *state reduction* in analogy with the R processes introduced by English theoretical physicist Roger Penrose of Oxford University [66, 67]. Each RS process is a self-connection of the timeless three-dimensional quantum vacuum. The appearance of a massive particle corresponding to a certain RS process taking place in a given region of the ordinary space-time derives from an opportune change of the quantum vacuum energy density.

The probability of the occurrence of a creation/destruction event for a quantum particle in a given point of space is linked with the opportune probability amplitudes for creation events and for destruction events. These probability amplitudes satisfy a time-symmetric extension of Klein-Gordon quantum relativistic equation[14]. At the non-relativistic limit, this general equation becomes a pair of Schrödinger equations, one for creation events, and the other for destruction events. On the basis of these equations, there are no true causal propagations from the future, in other words there is no possibility to have time travels.

In this approach, the three-dimensional quantum vacuum can be seen not only as an eigenstate of minimum energy, but also the network of all the possible RS processes of the field modes in an "undivided oneness", and it has to be regarded as a radical non-local and event-symmetric state. The three-dimensional quantum vacuum constrains and conveys the dynamical processes we observe. On this basis one can think that in the future the Bell's theorem not only will tell us the limits of hidden variables theories, but will be the cornerstone of a theory able to explain non locality as a residual effect, in particular conditions, from the manifestations of the primeval three-dimensional quantum vacuum. The "laboratory" non-locality appears as a particular case of this a-temporal wholeness.

In the mathematical formalism of the three-dimensional quantum vacuum model proposed by the authors, a quantum potential of the three-dimensional quantum vacuum at two components can be considered as the fundamental

---

[14] In relativistic quantum mechanics, the Klein-Gordon equation is the fundamental equation governing the behaviour of spinless particles.

mathematical element which guides the occurring of the processes of creation and annihilation of quanta in the different regions of the three-dimensional quantum vacuum. The equations regarding the quantum potential of the vacuum show that, both in the relativistic domain and in the non-relativistic domain, the quantum potential of the vacuum has a like-space, non-local, instantaneous action. Both for the processes of creation and for the processes of annihilation the quantum potential have a non-local, instantaneous action. The first component of the quantum potential regards the processes of creation of quanta, the second component – which is characterized by an opposed sign with respect to the first component – regards the processes of annihilation of quanta in the three-dimensional quantum vacuum. Moreover, according to the authors, the opposed sign of the second component with respect to the first component translates from the mathematical point of view the idea that, in the three-dimensional quantum vacuum, time exists only as a measuring system of the numerical order of material changes: the sign of the second component seems to indicate that it is not possible to go backwards in the physical time intended as a numerical order.

The connection between RS processes is possible because these events derive from the change of the energy density of the same timeless three-dimensional quantum vacuum (and the quantum potential can be considered the ultimate mathematical entity which guides the evolution of the occurring of the processes of creation or annihilation of quanta in the different regions of the three-dimensional quantum vacuum). According to this approach, therefore, the history of the universe, considered at the basic level, is given neither by the application of forward causal laws at initial conditions nor by the application of backward causal laws at final conditions. Instead, it is assigned as a whole as a complete network of RS processes that take place in the timeless three-dimensional quantum vacuum. Causal laws are only rules of coherence which must be verified by the network and are per se indifferent to the arrow of the time that we as observers seem to perceive in our level of physical reality.

The philosophy of this model as regards the link between our level of physical reality and the three-dimensional quantum vacuum actually follows someway the program that Bohm had already sketched out when dealing with clarifying the relationship between implicate and explicate order. Bohm's theory of holomovement between explicate and implicate order provides the description of the relation between the quantum vacuum and its manifestations. It is the passage from the timeless non-local background to the mathematical concepts of time and local in a continuous feedback. It is important to understand that holomovement does not happen in a time (intended as primary physical reality) but gives origin to the mathematical space-time itself by the activity of the RS processes and the changes of the energy density of the three-dimensional quantum vacuum associated with them.

1.12.1. Quantum Mechanics in the Timeless Three-Dimensional Quantum Vacuum

An interesting merit of this approach is represented by the possibility to obtain, on one hand, an interesting reading of the quantum dynamics as a particular aspect of such general theory and, on the other hand, a suggestive interpretation of gravity as a phenomenon emerging from the timeless three-dimensional quantum vacuum. On the basis of our approach of three-dimensional quantum vacuum, standard quantum mechanics might not be fundamental but rather emerge from a more fundamental descriptive level of reality, namely the three-dimensional quantum vacuum whose primary physical reality are RS processes which evolve according to appropriate equations. In fact, non-relativistic quantum mechanics based on the standard Schrödinger equation derives from the more general time-symmetric extension of Klein-Gordon quantum-relativistic equation in the case of creation events. Moreover, the standard quantum mechanical formalism in terms of the state vectors of Hilbert spaces emerges naturally by starting from the RS processes of the timeless three-dimensional quantum vacuum.

Let us consider a generic RS process which is constituted by creation of the quality ψ and destruction of the quality φ. These two events, which according to the three-dimensional timeless quantum vacuum model are the only manifest primary reality in the physical world, are correlated by opportune amplitude which can generally be developed as the sum of the partial amplitudes corresponding to different paths joining the two events. To be explicit, let us consider for example the well-known double-slit experiment. In such experiment the quality ψ is associated with the "preparation" of the "particle" in the source and the quality φ is associated with the "detection" of the "particle" behind the screen containing the slits. Indicating the two slits as f_1, f_2, for this process one obtains a total amplitude that is the sum of two partial amplitudes, one of which corresponds to a "passage through slit f_1", the other to a "passage through slit f_2". These two amplitudes interfere. Each of the two amplitudes is represented as the product of two free propagation amplitudes of the "particle": one corresponding to the propagation from the source to the slit, the other from the slit to the detector.

Hence: in this example of the double-slit experiment, either the "particle" interacts with the atoms which compose the screen, and is absorbed, or it does not interact with the screen at all. Only in this second case it can reach the detector, passing through f_1 or through f_2. The terms f_1, f_2 which appear in the expression of the total amplitude are therefore associated with interactions with the atoms of the screen (to be precise, "negative" interactions, i.e., non-interactions or absence of actual interactions). The correlation between the two events of creation of quality ψ and destruction of quality φ is therefore conditioned by the events characterizing the atoms which constitute the screen. The structure of the total amplitude includes these influences.

Here, the fundamental thing to emphasize is that the interaction events which appear in the total amplitude have not a physical reality of their own, they are virtual events appearing in the partial amplitudes as mere artifices of calculation. For example, the (non) interaction events f_1, f_2 in the example of the double-slit experiment are virtual events. By this we mean to say that these events are not pairs of emission/re-absorption of qualities by the physical vacuum, whereas the extreme events corresponding to the creation of quality ψ and the destruction of quality φ are. However, virtual events can become real. For example, a "particle" counter can be placed behind f_1 or f_2. In this case, though, one is no longer dealing with the original RS process, rather with other entirely different processes, for example the process represented by creation of the particle in f_2 and destruction of the particle in f_1.

Energy appears from the timeless three-dimensional quantum vacuum at the time of creation, and vanishes into the vacuum at the time of destruction, as do, in fact, the other quantities which make up the quantum vacuum. From a merely numerical standpoint, one can view creation as the release by the vacuum of a given amount of energy to the manifest physical world. Destruction will thus be the release by the manifest physical world of this amount of energy to the vacuum. Now, a system's exchanges of energy with the "outside" must be taken with a different sign according to their direction. In this regard, one may have the following two different situations:

- First situation. The energy released by the timeless three-dimensional quantum vacuum is positive and that absorbed by the vacuum is negative.

In this case, we shall have the exchange of an energy $E > 0$ at creation events, and the exchange of an energy $-E < 0$ at destruction events.

- Second situation. The energy released by the timeless three-dimensional quantum vacuum is negative and that absorbed by the vacuum is positive. In this case, we shall have the exchange of an energy $E < 0$ at creation events, and the exchange of an energy $-E > 0$ at destruction events.

The first situation corresponds to the forward propagation in time of a quantum of energy $E > 0$, or to the propagation backwards in time of a quantum of energy $E < 0$. The second situation corresponds to the forward propagation in time of a quantum of energy $E < 0$, or to the propagation backwards in time of a quantum of energy $E > 0$. In the timeless three-dimensional quantum vacuum model, both these situations are physically possible and are incompatible. If the first situation occurs, retropropagations in time of quanta with positive energy cannot exist; if the second situation occurs, retropropagations in time of quanta with negative energy cannot exist. This means that the direction of time ordering we perceive in our everyday life cannot be considered as a primary physical reality because it emerges from fundamental processes of a timeless three-dimensional quantum vacuum where time arrow has not an existence on its own, since both the situations illustrated above may occur. Time, in our everyday life, flows in one direction only while the RS processes – the most elementary processes of the physical world – are symmetrical in time.

In the model of the timeless three-dimensional quantum vacuum suggested by the authors, each elementary particle which emerges from an elementary RS process has a corpuscular and a wave nature. For example electron is a particle which moves through a three-dimensional quantum vacuum and it can be associated with an opportune wave of the vacuum which evolves according to the general equation of the three-dimensional quantum vacuum model. More precisely, the behaviour of a quantum particle as we know it from ordinary quantum mechanics derives from the equations of the timeless three-dimensional quantum vacuum model regarding the creation events. In the mathematical formalism developed by the authors of this book, the following interpretation of subatomic particles becomes permissible: in our world electrons and other elementary particles have precise positions at every time and follow precise trajectories which emerge from the evolution laws regarding creation events and the corresponding quantum potentials indicate the quantum force exerting on the corpuscle, which guides the corpuscle in the regions where the wave function of the vacuum is more intense. Therefore, according to the timeless three-dimensional quantum vacuum model of the authors, the quantum potential of ordinary non-relativistic quantum mechanics can be considered a consequence of a more fundamental quantum potential of the three-dimensional quantum vacuum (and an analogous result regards the quantum potential of Klein-Gordon's relativistic quantum mechanics).

Moreover, as regards the quantum potential of the three-dimensional quantum vacuum, it is important to emphasize that it must not be considered as an external entity in the vacuum but as an entity which contains an information about the background, namely as an entity which represents the three-dimensional quantum vacuum. In other words, the quantum potential of the vacuum can be considered a geometric entity of the vacuum, the information determined by the quantum potential of the vacuum is a type of geometrodynamic information "woven" into the quantum vacuum. The quantum potential of the vacuum has a geometric nature just because has a contextual nature, contains a global information on the environment in which the experiment (deriving from opportune RS processes of creation or destruction of opportune qualities) is performed; and at the same time it is a dynamical entity just because its information about the RS processes and their environment is active, determines the behaviour of the quantum particles (created or destroyed in the RS processes). On the basis of its mathematical expression, the action of the quantum potential of the vacuum is like-space, namely creates onto the particles (created or destroyed in the RS processes) a non-local, instantaneous action. In a double-slit experiment, for example, if one of the two slits is closed the quantum potential of the vacuum changes, and this information arrives instantaneously to the particle, which behaves as a consequence. This means that, at a fundamental level, the timeless three-dimensional quantum vacuum, through its special state represented by the quantum potential of the vacuum, acts as an immediate information medium in determining the motion of a subatomic particle.

In accordance with the ontology which derives from the three-dimensional quantum vacuum model here illustrated, the primary physical reality is constituted solely by RS processes. The causal connection between RS processes is not ensured by the propagation of an object in space-time intended as a primary physical reality, be it a wave, a corpuscle or any other entity which acts as an intermediary; rather, it is ensured by the common emergence of RS processes from the same timeless underlying three-dimensional quantum vacuum (and the non-local action of the quantum potential of the vacuum which is responsible of the evolution of the occurring of the creation and destruction events is instead the ultimate mathematical reality). By using a Bohmian terminology, we can define the reality constituted by the RS processes as the background from whose differentiation the foreground constituted by the events and their evolution (governed by the waves of the vacuum through the non-local action of the quantum potential) emerges. By attributing the status of primary physical reality solely to RS processes it follows that the space-time coordinates are labels associated with these processes and which express certain properties of relation between them. Space-time as such is only the domain setoff this labels and has no physical reality of its own. It is, so to speak, materialized by RS processes. In analogous way, also the background associated with the quantum potential is materialized by RS processes.

1.12.2. Gravitation as an Immediate Physical Phenomenon

In general relativity the gravitational interaction is the result of curvature of space-time. Stellar objects change geometry of space-time. Massive objects move in the direction of higher curvature of space-time. Bigger is the mass of a stellar object, more space-time is curved, bigger is the gravitational field.

However, this approach leads to well known problems if one aims to find a unifying picture which takes into account some basic aspects of the quantum theory. For this reason, several authors advocated different ways in order to treat gravitational interaction, in which the space-time manifold can be considered as an emergence of the deepest processes situated at the fundamental level of quantum gravity. In this regard, for example, in the second half of the 60s the Russian physicist Andrei Sacharov proposed to interpret gravity as "metric elasticity" of space [68]. Since 1994 the American physicists Bernard Haisch of California Institute for Physics and Astrophysics and Alfonso Rueda of California State University developed a model in which inertial mass and gravitational mass are interpreted as effects of the electromagnetic quantum vacuum [69, 70]. The American physicist Harold E. Puthoff of the Institute of Advanced Studies at Austin, Texas, suggested a polarizable vacuum model of gravitation in which the variation in the dielectric properties of the vacuum near massive objects determines effects on light propagation which are approximately equivalent to general-relativistic space-time curvature, as long as the fields are sufficiently weak [71]. Finally, more recently, the Italian physicist Maurizio Consoli of the Catania University constructed a model on ultra-weak excitations in a condensed manifold in order to describe gravitation [72, 73]. In Consoli's model, the physical vacuum is represented by a superfluid medium – a Bose condensate of elementary spinless quanta – whose long-range fluctuations, on a coarse-grained scale, resemble the Newtonian potential, allowing us thus to obtain the first approximation to the metric structure of classical general relativity. Under the construction of all of these models there is probably one underlying fundamental observation: as light in Euclid space deviates from a straight line in a medium with variable density, an "effective" curvature might originate from the same physical flat-space vacuum.

According to the model developed by the authors of this book, the timeless three-dimensional quantum vacuum whose most fundamental physical properties are its energy density and state-reduction (RS) processes (of creation-annihilation of quantum particles) can be considered the fundamental arena of gravity. In this picture, gravitation emerges at a fundamental level as an immediate physical phenomenon associated with the behaviour of the quantum vacuum energy density. The model of gravitation proposed by the authors introduces a three-dimensional background, a three-dimensional physical vacuum without the need to postulate the consideration of other processes (involving other more fundamental particles like Higgs particles): it is based only on the Planckian metric which leads directly to the definition of a Planck energy density as the maximum density that can be possibly sustained in space and on RS processes (of creation-annihilation of quanta). However, in this picture – just like, for example, in Consoli's model – there is at the basis the same underlying consideration: an "effective" curvature might originate from the same physical flat-space vacuum, from the same physical flat-space background.

In the model of a timeless three-dimensional quantum vacuum developed by the authors, the origin of gravitational phenomena lies simply in the changes of the quantum vacuum energy density associated with the RS processes. Here, the Planck energy density can be considered as the ground state of the same physical flat-space background; the appearance of material objects and subatomic particles derive from opportune RS processes of creation and correspond to opportune changes of the quantum vacuum energy density and thus can be interpreted as the excited states of the same physical flat-space background represented by the timeless three-dimensional quantum vacuum.

Gravity is a phenomenon determined by the changes of the quantum vacuum energy density and thus by the excited states of the same flat quantum vacuum. Each material particle endowed with mass is produced by a change of the energy density of quantum vacuum. Each body or particle diminishes quantum vacuum energy density. We measure this diminishing of quantum vacuum energy density as "inertial mass" and as "gravitational mass". When a material object (determined by an excited state of the three-dimensional quantum vacuum) moves, the diminishing of the quantum vacuum energy density causes a shadowing of the gravitational space which determines the motion of other material objects present in the region under consideration. The shadowing of the gravitational space is linked with a factor which represents the relatively small amount of the altered permittivity of the free space (with respect to the situation in which the quantum vacuum energy density coincides with the Planck energy density) and on a field associated itself with the changes of the quantum vacuum energy density (corresponding to the excited state under consideration of the quantum vacuum).

In the approach suggested by the authors, the quantum vacuum energy density can be considered the fundamental element that determines the gravitational interaction between material objects: gravity is carried by the effective curvature caused by a variable quantum vacuum energy density. And, at the same time, the equations of this model allow us to obtain results that are coherent and compatible with general theory of relativity.

A crucial result inside this model is that gravitational interaction is an immediate, direct phenomenon: the three-dimensional quantum vacuum acts as a direct medium of gravity. No movement of particle-wave is needed for its acting: gravity is transmitted directly by the quantum vacuum energy density, gravity is the result of the geometrical shape of the quantum vacuum energy density. No time is needed to transmit the information from a material object to the surrounding region in order to shadow the background and, on the other hand, no time is needed to transmit the information from the shadowed background to another material object in order to cause its movement.

One can also say that the gravitational space is a dynamic energy system where fluctuation of elementary constituents of space follows motion of stellar objects and elementary particles. Where energy density of quantum vacuum is less, space is more "stretched" and has a stronger tendency to "shrink". This "shrinking" force corresponds to the action of the gravitational field.

According to the view introduced by the authors, gravitational interaction is a "short distance interaction". Gravity does not work directly between stellar objects, it works in space in which stellar objects are existing: it is tied to the changes of the quantum vacuum energy density. In virtue of opportune values of the quantum vacuum energy density, for example, the region of space around the sun has tendency to shrink and this is the gravitational force that pulls the planets towards the sun.

The curvature of space-time which arises from the changes of the quantum vacuum energy density acts as a direct information medium that generates the motion of material objects in a gravitational field. Just like the information between two quantum particles in "Einstein-Podolski-Rosen" experiment does not travel between these two particles but is transmitted by a background space which is timeless and functions as an immediate information medium, in analogous way the gravitational interaction between two material objects is not linked with particles or waves which travel between them but is transmitted directly by the timeless space through its fundamental property represented by the quantum vacuum energy density.

The gravitational interaction between two massive bodies can be therefore interpreted as an immediate physical phenomenon. One can say that at a fundamental level the three-dimensional quantum vacuum can be considered as a direct medium that generates gravitational motion of material objects into direction of higher curvature of space-time and lower energy density of quantum vacuum. There is no direct attraction force between material objects. Presence of mass is diminishing quantum vacuum energy density. Diminished quantum vacuum energy density increases the curvature of space-time. The increasing of the curvature of space-time determined by the diminishing of the quantum vacuum energy density causes gravitational motion. Gravitational interaction between two masses is immediate: presence of one mass diminishes quantum vacuum energy density characterizing the region into consideration, this diminishing of the quantum vacuum energy density consequently increases curvature of space-time and this fact causes gravitational motion of the second mass. The first mass acts on the other mass indirectly via curvature of space-time determined by a variable quantum vacuum energy density.

This timeless interpretation of gravitation as an immediate physical phenomenon transmitted directly by the changes of the quantum vacuum energy density implies that, for example, the elapsed clock run for gravitational interaction between the earth and the sun is zero, that the gravitational interaction between earth and sun is immediate in the sense that the curvature of space-time caused by the presence of the sun (and thus by the diminishing of the quantum vacuum energy density around the sun) acts instantaneously on the earth determining its motion in its own trajectory. The sun acts instantly on the earth via the variable quantum vacuum energy density (determined by the sun) which functions as a direct, immediate information medium between the earth and the sun.

In original papers of 1916 Einstein do not mention gravitational waves [74]. This idea arises few months later. Einstein introduces gravitational waves as space-time perturbations. With the introduction of gravitational waves that propagate with light speed gravity is interpreted as a non-immediate phenomenon as propagation of gravitational waves requires some "tick" of clocks.

The Italian physicist of the University of Milan Alberto Loinger considers that gravitational waves are only hypothetical and do not exist in the physical world: "The gravitational waves are non-physical sinuosities generated, in the last analysis, by undulating reference frames" [75-76]. In the 1960s, Joseph Weber began his experimental work to detect gravitational waves [77]. He was essentially alone in this field of research. Later, theoretical papers of Wheeler, Bondi, Landau and Lifshitz, Isaacson, Thorne and others, as well as experimental work of Weber, Braginski, Amaldi and others opened a new area of research in this field. However, gravitational waves have not yet been detected. As underlined by Schorn in the reference [78], "To search for gravitational waves in a laboratory, classical or quantum mechanical detectors can be used. Despite the experiments of Weber (1960 and 1969) and many others (Braginskij and others, 1972; Drever and others, 1973; Levine and Garwin, 1973; Tyson, 1973; Maischberger and others, 1991; Abramovici and others, 1992; and Abramovici and others, 1996) and theoretical calculations and estimations (Braginski and Rudenko, 1970; Harry and others, 1996; and Schutz, 1997), gravitational waves have never been observed directly in laboratory". Therefore, we can conclude that no experimental evidence exists until now against the interpretation of gravity as immediate physical phenomenon which arises from the quantum vacuum energy density.

1.12.3. Gravity and Quantum Behaviour as Two Aspects of the Timeless Three-Dimensional Quantum Vacuum

Finally, the timeless three-dimensional quantum vacuum model introduces the interesting perspective to interpret gravity and quantum behaviour as two different aspects of a same source. In this regard, it is interesting to observe that the Iranian physicists A. Shojai and F. Shojai of the Teheran school recently developed an interesting toy model which, by studying the behaviour of spinless particles in a curved space-time, demonstrates that the quantum potential contributes to the curvature that is added to the classic one and reveals deep and unexpected connections between gravity and the quantum phenomena. By the investigation of the coupling of purely gravitational effects and purely quantum effects of a particle in a general background space-time metric, this approach can obtain a fundamental equivalence between quantum effects of matter and a curved space-time. By the analysis of the quantum effects of matter in the framework of bohmian mechanics, A. Shojai's and F. Shojai's model shows that the motion of a spinless particle with quantum effects is equivalent to its motion in a curved space-time. The quantum effects of matter as well as the gravitational effects of matter have geometrical nature and are highly related: the quantum potential can be interpreted as the conformal degree of freedom of the space–time metric and its presence is equivalent to the curved space-time. In fact, the presence of the quantum force is just like having a curved space–time which is conformally flat and the conformal factor is expressed in terms of the quantum potential. The important conclusion of F. Shojai's and A. Shojai's model is that the presence of the quantum potential is equivalent to a curved space-time, providing thus a fundamental geometrization of the quantum aspects of matter [79, 80].

In analogy to the conclusions of F. Shojai's and A. Shojai's model, our model of a timeless three-dimensional quantum vacuum characterized by RS processes shows that quantum behaviour of matter and gravity constitute two different aspects deriving from the same timeless three-dimensional quantum vacuum. More precisely, by starting from Bohm's version of Klein-Gordon equation of the generic component of the probability amplitude of the occurrence of creation event for a quantum particle, the approach developed by the authors show that this generic component (and thus the quantum potential associated with it) has an important link with the curvature of the ordinary space-time we perceive. This means that the effects of gravity on geometry and the quantum effects on the geometry of space-time are highly coupled and that both of them derive from the RS processes of the timeless three-dimensional quantum vacuum. The important conclusion one can draw from the approach of the authors is that the presence of the quantum potential of the vacuum is equivalent to a curved space-time. On the basis of this model of the three-dimensional quantum vacuum, it becomes so permissible the following interpretation of the curvature of space-time (and thus of gravitation) inside the timeless three-dimensional quantum vacuum characterized by RS processes. RS processes associated with creation events of quantum particles determine a quantum potential of the vacuum which is equivalent to the curvature of the space-time. The quantum potential of the vacuum corresponding to the generic component of the probability amplitude of the creation event of a quantum particle is tightly linked with the curvature of the space-time we perceive. In other words, one can say that RS processes, through the manifestation of the quantum potential of the vacuum, lead to the generation, in our macroscopic level of reality, of a curvature of space-time and, at the same time, the space-time metric is linked with the quantum potential of the vacuum which influences and determines the behaviour of the particles (themselves corresponding with creation events from the timeless three-dimensional quantum vacuum).

In synthesis, the approach of the three-dimensional timeless quantum vacuum suggested by the authors achieves a significant geometrization of physics which has origin in the fundamental processes of the timeless three-dimensional quantum vacuum, namely the RS processes and the changes of the energy density corresponding to them. In this picture, one can say that the space-time geometry sometimes looks like what we call gravity and sometimes looks like what we understand as quantum behaviours and both these features of physical geometry emerge from the RS processes of the timeless three-dimensional quantum vacuum and the changes of the three-dimensional quantum vacuum energy density. In this approach, the fundamental result is obtained that the quantum behaviour of matter and gravity appear as two different aspects of the same timeless three-dimensional quantum vacuum.

1.13. Relativistic Effects of Relative Velocity of Material Change Start with Massive Particles

Special relativity and general relativity imply that at the scale of the photon and below at the Planck scale relativistic effects of relative velocity of material change do not exist. In all inertial systems velocity of a photon clock is unchanged. Change of gravity does not affect velocity of a photon clock.

Time dilatation starts by atomic clocks. Changes of velocity of atomic clocks on the orbit stations and on the surface of the earth are considered by Global Positioning System (GPS).

We have a "photon clock" made out of two mirrors A and B. Photon is moving from A to B, back to A and so on. One motion of the photon between A and B is a "tick" of the clock. Let us take two photon clocks. One photon clock is on the surface of the earth, the second is 4200 meters below at the bottom of the mine shaft. Velocity of light is invariant on gravity; both of clocks will "tick" with the same velocity.

Let us take two atomic clocks. One clock we put beside photon clock on the surface and second beside clock that is 4200 meters deep. According to the relativistic gravitational effect second atom clock will in 30 days "tick" faster as the atom clock on the surface for the quantity $\Delta t = 1.23 \cdot 10^{-6}$ seconds.

Photon clock will also have unchanged velocity in all different inertial systems, because velocity of light is equal in all different inertial systems. Similar experiments with two atomic clocks and two photon clocks can be carried out by putting one atomic clock and one photon clock in the orbit station and one atom clock and one photon clock on the surface of the earth. Photon clocks on the orbit station and on the surface of the earth will have the same velocity. Atomic clock on the orbit will "tick" with slower velocity than atomic clock on the surface of the earth.

1.13.1. "Gedanken experiment" with Atomic and Photon Clocks

By photon clock "time dilatation" is in contradiction with "length contraction". Let us consider a photon clock in a fast airplane. Mirrors are fixed in a way that photon moves along the direction of motion of the airplane. Distance between the mirrors is shortened by the length contraction. Because of the shorter distance between mirrors a path for the photon is shorter and so photon clock on the airplane "ticks" faster than same construction photon clock on the surface of the earth.

Experimental data show that atomic clock in a fast airplane ticks slower than atomic clock on the earth. Solution of this contradiction is in a proposition that photon clock in the airplane do not shorten. "Length contraction" is only a mathematical calculation that has no correspondence with the physical world. "Time dilatation" has also no correspondence in the physical world. What really happens by "time dilatation" is that velocity of material change (velocity of clocks including) slows down. Material change (clocks's run included) occurs in space only and not in time.

1.13.2. Constancy of Light Velocity is Connecting Special Relativity with General Relativity

Special relativity is postulated on constancy of light velocity. Equality of inertial and gravitational mass is connecting special relativity and general relativity. Discussing on possibility that gravity infects velocity of light would put under question connectedness of special relativity and general relativity. Gravitational red shift shows that gravity influences only frequency and not velocity of light. This means that "thought experiment" with photonic clock under the surface of the earth is correct: velocity of photonic clock is invariant on gravity. In stronger gravity photon moving between mirror A and B change only frequency and not velocity. And this means that relativistic gravitational effect of relative velocity of material change starts with massive particles.

"Gedanken experiment" with atomic and photon clocks shows that invariance of light velocity excludes existence of relativistic effects of relative velocity of material change by photon. Relativistic effects of relative velocity of material change start with massive particles.

"Time dilatation" is a relativistic effect of diminishing of velocity of atomic clocks running in space. Time cannot shrink as time does not exist in the physical universe. Clocks run in space only and not in time.

1.14. Timeless Universe in Dynamic Equilibrium

Astronomical observations show that in the universe energy circulates in cycles. In black holes matter is disintegrating in quanta of empty space. In vast outer space between galaxies quanta of empty space are forming “cosmic rain” – subatomic particles that get structured into matter. This circulation of the energy “matter-space-matter” has no beginning and no end [81].

According to the first law of thermodynamics energy can not be created and can not be destroyed. Energy of the universe is a non-created entity; it can only change its forms. Universe is a non-created system in a permanent dynamic equilibrium. In the universe we have three basic types of energy: energy of matter (E_m), electromagnetic energy (E_{em}) and energy of empty space (E_s). On the basis of the treatment made in chapter 1.12, energy of matter increases with the decreasing of the quantum vacuum energy density; instead energy of empty space decreases with the increasing of the amount of mass characterizing the region into consideration and thus increases with the increasing of the quantum vacuum energy density. Therefore, where energy of matter is high, energy of empty space is low. Electromagnetic energy can be defined as the form of energy which arises from electromagnetic radiations. In virtue of the famous Einstein’s equation regarding the equivalence between mass and energy, we can draw that there is an equivalence between energy of matter and electromagnetic energy, that energy of matter and electromagnetic energy can also be transformed into each other.

Our idea is that also energy of matter E_m and energy of space E_s can be transformed into each other. In black holes energy of matter E_m is transforming into energy of space E_s. Energy of matter and energy of empty space (and thus energy of the quantum vacuum) are in dynamic relation: presence of matter in space diminishes energy density of space. In the centre of a black hole matter is extremely dense and energy density of empty space is extremely low. Energy of matter is transformed into electromagnetic energy and this further into energy of space:

$$blackhole: E_m \to E_{em} \to E_s .$$

Absence of matter increases energy density of empty space. In vast outer space between galaxies space is extremely dense. Quanta of space are forming “cosmic rain”. Well known researcher on cosmic rain is American physicist

Michael W. Friedlander [82]. According to our view the "cosmic rain" is a physical phenomenon where energy of empty space E_s is transforming into electromagnetic energy E_{em} which is then structuring into subatomic particles which build matter:

$$outer \quad space: E_s \rightarrow E_{em} \rightarrow E_m \cdot$$

This circulation of the energy "$E_m \rightarrow E_{em} \rightarrow E_s \rightarrow E_{em} \rightarrow E_m$" has no beginning and no end. Energy cannot be destroyed and cannot be created, only changes its forms. Universe is a non-created system in a permanent dynamic equilibrium. In general theory of relativity presence of matter changes the geometrical structure of space; it increases its curvature. As we have said in chapter 1.12, our idea is that curvature of space-time depends on energy density of the three-dimensional quantum vacuum. Less is the quantum vacuum energy density, higher is the curvature. As a consequence, in regions of "outer space" between galaxies space is extremely dense and almost flat.

The cosmological view here suggested implies that universe is a non created reality rejuvenating itself continuously. Universe is "young for ever". Idea of "creation" is borrowed from religions. Creation implies "creator" which is behind the universe, super-natural. Creator has attributes of God. Here we suggest the idea that universe is a self-renovating system and functions perfectly without God. There is no God behind the universe. Universe is God itself.

According to the Big Bang Theory universe started with a huge explosion and is continuously expanding. This view has several unanswered questions.

Defenders of Big Bang say that in the first moments there was an "inflation phase", where energy of matter and energy of space (also called gravitational energy) were created practically out of nothing. In his book *The Brief History of Time* Stephen Hawking describes inflation phase as multiplication of energy of matter and gravitational energy [83]. Here we interpret Hawking's gravitational energy as energy of the empty space. Hawking says that the sum of both energies (namely energy of matter and gravitational energy) in the universe is always zero (see also the book [84] for a review of these concepts). In inflation phase of the first moments after big bang both energies are multiplied in the same way as it is in mathematics:

1 + (-1) = 0
2 + (-2) = 0
n + (-n) = 0

These mathematical equations above does not answer question of how energy in Big Bang is created. Physics and cosmology can not be explained with mathematics. Mathematics is only a description tool of the world; it is not its explicator. According to our view there was no creation of the universe and no creation of energy of matter and space. Energy of the universe is non-created, eternal. It flows in a cyclic numerical order as following:

$$E_m \rightarrow E_{em} \rightarrow E_s \rightarrow E_{em} \rightarrow E_m .$$

Moreover, it is important to underline that idea of expanding universe is based on "red shift". We observe that light coming from far galaxies has lower frequency. This is the so called "Doppler effect" by the light. Frequency of sound and frequency of light which are coming from the sources that move away from us are lower than frequencies from the still sources where distance between us and the source are not increasing. Calculations done on the red shift confirms that around 60% of red shift is due to the lost of energy when light is "pulling out" of strong gravity field of the far galaxies. Expanding of the universe is still under question and we do not have a final answer yet.

The trouble with expanding universe is that it does not answer the question from where the energy of the expanding space is coming: if universal space is increasing from where is energy of space created?

Big Bang theory has no answer on this question. On the other hand in the view of the universe in dynamic equilibrium theory creation of energy is not a problem at all. Energy is non-created; it only changes its forms.

Moreover, it is important to point out that the view according to which in the universe there is a continuous transformation of matter into space and vice versa implies that the increasing of entropy of matter must be considered only a temporary fact: the purpose of the universe is indeed to renovate "old matter" endowed with high entropy with "new matter" endowed with low entropy. This approach can put thus under discussion the standard point of view of mainstream science: if universe is a system that recreates itself continuously, it cannot be considered as a closed system and we cannot state how big it is. The model of universe in dynamic equilibrium seems to be able to reproduce all the predictions of the standard cosmological model with the same accurateness and has a supplementary predictive power to face many questions that the standard model cannot face: what happened in the initial singularity? What is the final destiny of the universe? In this model, universe is without end, is timeless. While big bang theory has a strong dependence on precise initial conditions, namely presents the so-called problem of the beginning of the universe, instead our model of universe in dynamic equilibrium must not face the problem of the beginning. Our view, in fact, foresees no initial configuration of the universe. Time exists only as a numerical order of material changes and changes happen for ever, and also space is without end. Our three spatial dimensions remain infinite during the evolution of the universe. By means of the formation of empty quanta of space in black holes and the formation of new matter in the outer space, universe takes new energy and cycles of transformation space-matter-space-matter-… are possible. These cyclic processes happen periodically without an end.

Moreover, it is important to underline that our model seems coherent also in virtue of some recent astronomical observations. Before all, some observations made by astronomers of Virginia University and Massachusetts University show that the active galactic nucleus of Milky Way would be eating Sagittarius galaxy and, at the same time, would emit fresh gas. These results are difficultly explainable in the big bang model while are easily reproducible inside the model of the universe in dynamic equilibrium. Secondly, astronomical observations put in evidence a diminishing of the mass of neutron stars. Also this phenomenon is not explainable in the standard model; instead, inside the model of a timeless universe in dynamic equilibrium, it can be explained in a coherent way. It is linked to the fact that in neutron stars matter is transforming into energy of empty space.

REFERENCES

[1] Góźdz, A.; Stefanska, K. *Jour Phys: Conf Ser* 2008, *104*, 012007.

[2] Sorli, A.; Fiscaletti, D.; Klinar, D. *Phys Ess* 2010, *23*, 330-332.

[3] Sorli, A.; Fiscaletti, D.; Klinar, D. *Phys Ess* 2011, *24*, 11-15.

[4] Mach, E. *Die Mechanik in ihrer Entwicklung historisch-kritsch dargestellt*; Barth: Leipzig, DE, 1883; English translation: *The Science of Mechanics*; Open Court: Chicago, IL, 1960.

[5] Yourgrau, P. *A World Without Time: The Forgotten Legacy of Gödel And Einstein*; Basic Books: New York, NY, 2006.

[6] Barbour, J.; Bertotti, B. *Proc Royal Soc A* 1982, *382*, 295–306.

[7] Barbour, J. *Class Quant Grav* 1994, *11*, 2853-2873.

[8] Barbour, J.; Foster, B. Z.; Murchadha, N. O' *Class Quant Grav* 2002, *19*, 3217–3248; e-print arXiv:gr-qc/0012089.

[9] Woodward, J. F. *Found Phys Lett* 1996, *9*, 1-23.

[10] Rovelli, C. *Phys Rev D* 1991, *43*, 442-456.

[11] Rovelli, C. *Nuovo Cimento* 1995, *110B*, 81-93.

[12] Rovelli, C. In *Physics meets philosophy at the Planck scale*; Callender, C.; Huggett, N.; Eds.; Cambridge University Press: Cambridge, MA, 2001.

[13] Rovelli, C. *Phys World* 2003, *7*, 1-5.

[14] Rovelli, C. *Quantum gravity*; Cambridge University Press: Cambridge, MA, 2004.

[15] Barbour, J. (2009). The Nature of Time. http://arxiv.org/abs/0903.3489

[16] Caticha, A. (2010). Entropic dynamics, time and quantum theory. arXiv:1005.2357v3 [quant-ph]

[17] Palmer, T. N. (2009). The Invariant Set Hypothesis: A New Geometric Framework for the Foundations of Quantum Theory and the Role Played by Gravity. http://arxiv.org/abs/0812.1148

[18] Girelli, F.; Liberati, S.; Sindoni, L. (2009). Is the notion of time really fundamental? http://arxiv.org/abs/0903.4876

[19] Prati, E. (2009). The nature of time: from a timeless hamiltonian framework to clock time metrology. arXiv:0907.1707v1

[20] Licata, I. "Dinamica reticolare dello spazio-tempo", In *Inediti Scientifici Andromeda*, 27; Bologna, IT, 1989.

[21] Licata, I. *Hadronic Journal* 1991, *14*, 225-250.

[22] Sorli, A.; Sorli, I. K. *Front Persp* 2005, *14*, 38-40.

[23] Sorli, A.; Fiscaletti, D. *Phys Ess* 2012, *25*, 141-143.

[24] Manaresi, R.; Selleri, F. *Found Phys Lett* 2004, *17*, 65-79.

[25] Selleri, F. "Space and time should be preferred to spacetime – 1", *International workshop Physics for the 21st century*, 5-9 June 2000.

[26] Selleri, F. "Space and time should be preferred to spacetime – 2", *International workshop Physics for the 21st century*, 5-9 June 2000.

[27] Duffy, M.; Levy J. eds. *Ether space-time & cosmology*; Apeiron: Montreal, CA, 2009, Vol. 3.

[28] Deutsch, D. *The Fabric of Reality*; Allen Lane The Penguin Press: London, UK, 1997; p. 263.

[29] Bohm, D. *Quantum theory*; Prentice-Hall: New York, NY, 1951.

[30] Fiscaletti, D.; Sorli, A. *Phys Ess* 2008, *21*, 245-251.

[31] Fiscaletti, D.; Sorli, A. *Prespacetime Journal* 2010, *1*, 883-898.

[32] Bohm, D. *Phys Rev* 1952, *85*, 166-193.

[33] Bohm, D. *Phys Rev* 1953, *89*, 458-466.

[34] Bohm, D. In *Symposium on the foundation of modern physics – 1987*; Lahti, P. and Mittelstaedt, P.; Eds.; World Scientific Publishing Co: Singapore, SI, 1988.
[35] Bell, J. S. *Physics* 1964, *1*, 195-200.
[36] Chew, G. *Sc Prog* 1960, *51*, 529-539.
[37] Bohm, D. *Wholeness and Implicate Order*; Routledge & Kegan Paul: London, UK, 1980.
[38] Bohm, D.; Hiley, B. J. *The undivided universe: an ontological interpretation of quantum theory*; Routledge: London, UK, 1993.
[39] Hiley, B. J. "Non-commutative geometry, the Bohm interpretation and the mind-matter relationship", in *Proc. CASYS'2000*, Liege, Belgium, Aug. 7-12, 2000.
[40] Fiscaletti, D. *Prospettive alla ricerca del graal. Verso una visione unitaria di spazio, materia e vita*; Aracne Editrice: Roma, IT, 2010.
[41] Sbitnev, V. I. *Kvantovaya Magiya* 2008, 5, 1101-1111; URL http://quantmagic.narod.ru/volumes/VOL512008/p1101.html.
[42] Sbitnev, V. I. *International Journal of Bifurcation and Chaos* 2009, *19*, 2335-2346; e-print arXiv:0808.1245v1 [quant-ph].
[43] Fiscaletti, D.; Sorli, A. *Annales UMCS Sectio AAA: Physica* 2012, *57*, 47-72.
[44] Wharton, K. B. *Found Phys* 2007, *37*, 159-168.
[45] Fiscaletti, D.; Sorli, A. *IUP Jour Phys* 2010, *3*, 34-49.
[46] Fiscaletti, D.; Sorli, A *IUP Jour Phys* 2011, *4*, 34-49.
[47] Eckle, P.; Pfeiffer, A. N.; Cirelli, C.; Staudte, A.; Dörner, R.; Muller, H. G.; Büttiker, M.; Keller, U. *Science* 2008, *322*, 1525-1529.
[48] Pirandola, S.; Braunstein, S. L.; Mancini S.; Lloyd S. *Eur Lett* 2008, *84*, 20013-1-20013-6.
[49] Hegerfeldt, G. C. *Phys Rev Lett* 1994, *72*, 596 – 599.
[50] Whong Chao Wou (2008). The Imaginary Time in the Tunneling Process. http://arxiv.org/abs/0804.0210.
[51] Devetak, I.; Shor, P. V. *Communications in Mathematical Physics* 2005, *256*, 287-303.
[52] Erickson, G. W. *Phys Rev Lett* 1971, *27*, 780-783.
[53] Jaffe, R. L. *Phys Rev D* 2005, *72*, 021301, 1-5; e-print http://arXiv:hep-th/0503158v1.
[54] Klimchitskaya, G. L.; Mohideen, U.; Mostepanenko, V. M. *Rev Mod Phys* 2009, *81*, 1827-1885.
[55] Beck, C.; Mackey, M. C. *Fluct Noise Lett* 2007, *7*, C27-C35.

[56] Bettini, A. *Introduction to elementary particle Physics*; Cambridge University Press: Cambridge, Ma, 2008.
[57] Rubakov, V. A *Phys–Usp. Adv Phys Sc* 2007, *50*, 390–396.
[58] Sahni, V. *Lect Not Phys* 2004, *653*, 141-180; e-print arxiv.org/abs/astro-ph/0403324v3.
[59] Chernin, A. D. *Phys–Usp. Adv Phys Sc* 2008, *51*, 253–282.
[60] Padmanabhan, T. "Darker side of the Universe", *29 International Cosmic Ray Conference Pune*, *10*, 47-62, 2005.
[61] Sorli, A. *Jour Adv Phys* 2012, *1*, 110-112.
[62] Chiatti, L. (2012). The transaction as a quantum concept. arXiv.org/pdf/1204.6636
[63] Licata, I. *Eur Phys Jour* 2013.
[64] Fiscaletti, D.; Sorli, A. "Perspectives about quantum mechanics in a model of a three-dimensional quantum vacuum where time is a mathematical dimension", in preparation, 2014.
[65] Fiscaletti, D.; Sorli, A. "Energy density of space and quantum behaviour", in preparation, 2014.
[66] Penrose, R. *The Emperor's New Mind*; Oxford University Press: Oxford, UK, 1989.
[67] Penrose, R., *The Road to Reality*; Oxford University Press: Oxford, UK, 2004.
[68] Sakharov, A. D. *Doklady Akad. Nauk S.S.S.R.* 1967, *177*, 70-71.
[69] Haisch, B.; Rueda, A.; Puthoff, H. E. *Phys. Rev A* 1994, *48*, 678-694.
[70] Rueda, A.; Haisch, B. (2005). Gravity and the quantum vacuum inertia hypothesis. arXiv:gr-qc0504061v3
[71] Puthoff, H. E. *Found Phys* 2002, *32*, 927-943.
[72] Consoli, M. (2009). Ultraweak excitations of the quantum vacuum as physical models of gravity. arXiv:0904.1272v2 [gr-qc]
[73] Consoli, M. In: Vision of oneness; Licata, I.; Sakaji, A.; Eds.; Aracne Editrice: Rome, IT, 2011.
[74] Einstein, A. *Sitz. Ber. Kon. Preus, Ak. Wiss* 1916, 688.
[75] Loinger, A. (1998). The Gravitational Waves are Fictitious Entities. http://xxx.lanl.gov/abs/astro-ph/9810137
[76] Loinger, A. (2004). The Gravitational Waves are Fictitious Entities-II. http://arxiv.org/vc/astro-ph/papers/9904/9904207v1.pdf
[77] Ciufolini, I.; Gorini, V. (2004). Gravitational Waves, Theory and Experiment (An Overview). http://bookmarkphysics.iop.org/fullbooks/0750307412/ciufoliniover.pdf
[78] Schorn, H-J. *Int Jour Theor Phys* 2001, *40*, 1427-1452.

[79] Shojai, A.; Shojai, F. *Phys Scr* 2001, *64*, 413-416.

[80] Shojai, F.; Shojai, A. (2004). Understanding quantum theory in terms of geometry. arXiv:gr-qc/0404102 v1

[81] Sorli, A.; Fiscaletti, D. *Electr Jour Theor Phys* 2005, 2, 7-13.

[82] Friedlander, M. W. *A Thin Cosmic Rain: Particles from Outer Space*; Harvard University Press: Cambridge, MA, 2000.

[83] Hawking, S. W. *A brief history of time*; Bantam books: New York, NY, 1998.

[84] Hawking, S. W.; Penrose R. *The Nature of Space and Time*; Princeton University Press: Princeton, UK, 1996.

Chapter 2

TOWARDS A UNIFIED VIEW OF THE PHYSICAL WORLD

ABSTRACT

The 20th century theoretical physics brought the idea of a unified vacuum as a fundamental medium subtending the observable forms of matter, energy and space-time. In the attempt to unify quantum theory and general relativity different approaches have been developed which invoke different mechanisms of description of the fundamental level of physical reality. In particular, the idea that time exists only as a numerical order of material changes introduces interesting perspectives towards a real timeless "theory of everything" inside a fundamental timeless granular space. In this chapter, after analysing the fundamental features of some relevant unitary theories of contemporary physics (such as superstring theory and loop quantum gravity) and making some considerations about being versus becoming and atomism versus monism in modelling physical reality at a fundamental level, we will focus our attention on the timeless "theory of everything" suggested by the authors of this book in some recent articles. Despite the ideas of this model are mostly speculative, we will show that they can open interesting perspectives as regards the interpretation of some important physical phenomena. We will enunciate the postulates of our "theory of everything" and we will see the interpretation of the main physical entities (in particular of subatomic particles and their interactions) inside this approach. Then, in the final part of the chapter, we will show that the replacement of the concept of time with the numerical order of material change allows us to provide a new interesting interpretation of superstring theory and loop quantum gravity. In particular, we will show that our timeless theory of everything can go beyond the results of superstring theory and of loop quantum gravity, can provide a significant unification of the fundamental ideas of superstring theory and of loop quantum gravity (as well as of another suggestive approach, the theory of the Wave Structure of Matter, which can resolve the paradoxes of quantum theory).

2.1. Towards a Deeper Structure of Subquantum Physics as a Fundamental Medium of the Universe

The concept of a physical field that subtends the three-dimensional background of space as well as time emerged in the course of the 20th century. Thanks to the famous Michelson-Morley experiments the ether was removed from physics and the absolute vacuum took its place. However, the vacuum turned out to be far from empty space. In the "grand unified theories" (GUTs) developed in the second half of last century the concept of the vacuum transformed from empty space into the medium that carries the zero-point field, or ZPF, where energies turn out to be present even when all classical forms of energy vanish (namely at the absolute zero of temperature). Even more interactions have come to light between this fundamental field and observed things and processes. In the 1960s Dirac showed that fluctuations in fermionic fields produce a polarization of the ZPF of the vacuum, whereby the vacuum in turn affects the particles' mass, charge, spin, or angular momentum [1]. At around the same time Andrei Sakharov proposed that relativistic phenomena (the slowing down of clocks and the shrinking of yardsticks near the speed of light) are the result of effects induced in the vacuum due to the shielding of the zero-point field by charged particles [2]. In current super-grand unified theories (super-GUTs) all the forces and fields of the universe arise from a "unified vacuum".

On the basis of the quantum field theories which describe the known particles and their interactions, the unified vacuum turns out to be characterized by different levels or domains, in other words there are various contributions to the energy density of the unified vacuum. As regards the description of the geometry underlying quantum processes, the so-called Standard Model of particle physics – which developed since the 70s of the last century and had success thanks to the experiments performed at the beginning of the 80s – is the fundamental theoretical framework that integrates our current knowledge of the subatomic world and its fundamental interactions (namely, electromagnetism, weak force and strong force)[1] and thus which constitutes the foundation for the characterization of the

[1] Electromagnetic forces act between particles endowed with an electric charge; today they are explained by a theory, known as quantum electrodynamics, according to which the transmitters of

quantum vacuum. On the basis of the Standard Model, one can infer that the total vacuum energy density has at least the following three contributions: the fluctuations characterizing the zero-point field, the fluctuations characterizing the quantum chromodynamic level of subnuclear physics (of quarks[2] and gluons[3]) and the fluctuations linked with the Higgs field[4]. Moreover, one can speculate that there are also contributions from possible existing sources outside the Standard Model of particle physics (for instance, GUT's, string theories, or other unifying field theories). There is no structure within the Standard Model which suggests any relations between the different contributions to the total vacuum energy density here mentioned, and it is therefore permissible to assume that the total vacuum energy density is, at least, as large as any of the individual contributions. In order to reconcile the vacuum energy density estimate within the Standard Model with the observational limits on the cosmological constant $|\Lambda| < 10^{-56} cm^{-2}$, the usual programme is to "fine-tune" [3].

As regards the role of the different contributions to the vacuum energy density, in the recent paper *Physical vacuum as a system manifesting itself on various scales – from nuclear physics to cosmology*, the Russian physicist Timashev examined the suggestive possibility of considering the physical vacuum as a unified system governing the processes taking place in microphysics and macrophysics, which manifests itself on all space-time scales, as a substance that unites and forms all the known types of interaction – strong, electromagnetic, weak and gravitational [4].

Current research exploring the behaviour of subatomic particles and their interactions suggests therefore that the unified vacuum subtending the observable forms of matter, energy and space-time must be considered not only as a theoretical artefact, a requirement of the mathematics of the field theory, but really as a part of physical reality, as a fundamental medium in the universe [5, 6]. The evidence for the realistic concept of vacuum is of two kinds. On one hand,

this interaction are photons. Strong interaction link protons and neutrons inside atomic nucleus (and, in another context, is the interaction that exert between quarks); it is very intense and its range is very small (about 10^{-35} meters). Weak interaction acts by disintegrating some unstable particles.

2 Quarks are the fundamental constituents of particles like protons and neutrons and of other particles which interact through strong interactions.

3 Quantum chromodynamics implies that the forces of interaction between quarks is transmitted by a new entity of particles, the gluons, of eight different species. As regards the interactions between quarks, known also as color forces, the gluons have the same role that the photons have as regards the electromagnetic interactions.

4 The Higgs field is the field that in the Standard Model is introduced in order to explain the origin of the mass of all the particles.

more and more scientists are coming to the conclusion that the level of quanta, and of space-time itself, is not the ultimate level in the universe. There is also a level below quanta and space-time, a level from which space-time and the quanta that populate it have emerged. The other kind of evidence concerns the observation that quanta and the things composed of quanta (living organisms included) are intrinsically and "non-locally" connected. This suggests the possibility that the fundamental level of the universe is not merely at the origin of the things that populate space and time, but is also the medium that interconnects them. In other words, it seems that non-locality has to be introduced ab initio as ultimate principle of the fundamental level of reality, of the unified vacuum that constitutes the basis of all physical phenomena.

The realistic concept of the vacuum completes and complements Einstein's theory of relativity (although it places in doubt one of its pillars, the constancy of the speed of light). Relativity theory views space-time as relative and dynamic, interacting with matter and energy. In relativity space-time is the "background" against which the events of the manifest world unfold. But the origins of this background are not accounted for in relativity theory: space-time is simply "given," together with matter and energy. This is much the same in the currently elaborated theories of everything (TOE's). The current TOEs would be truly theories of everything only if they were "background independent"; that is, if they did not merely assume the presence of space-time, but showed how it arose in the universe. The TOEs developed to date, based for the most part on string and super-string theories, cannot do this. The majority of the versions of strings works, just like quantum field theory, which is its closest relative providing the most general syntax we know to describe forces (and thus like also the Standard Model of particle physics), with a flat Minkowski space-time, while a correct, authentically relativistic (in the sense of general relativity) theory should be independent on the background, or not presuppose any metric signature.

The current impasse of theoretical physics points towards the necessity to recognize a deeper – and non-local – floor of the universe. "If we are ever going to find an element in nature that explains space and time," Princeton physicist John Archibald Wheeler asserted, "we surely have to find something that is deeper than space or time – something that itself has no location in space or time." [7].

There is also independent evidence speaking to the reality of a cosmic plenum. Since Einstein published his general theory of relativity in 1915, important evidence has come to light regarding the existence of a medium that would underlie the observable manifestations of the universe. Initially this cosmic medium was identified with space itself. In the nineteenth century, mathematician

William Clifford suggested that small portions of space are analogous to little hills on a surface that is, on average, flat; the ordinary forces of geometry do not hold for them [8]. The property of space to be curved or distorted, he said, is continually being passed on from one portion of space to another after the manner of a wave. This variation in the curvature of space is what really happens when matter moves. Thus in the physical world nothing else takes place but this wavelike variation.

In his 1930 paper *The Concept of Space* Einstein himself noted, “We have now come to the conclusion that space is the primary thing and matter only secondary; we may say that space, in revenge for its former inferior position, is now eating up matter.” [9]. A few years following the publication of Einstein’s thought, Erwin Schrödinger restated this basic insight: “What we observe as material bodies and forces” he noted, “are nothing but shapes and variations in the structure of space.” [10].

More and more theories ascribe physical properties to space, more exactly, to the field or medium that subtends space. In particular, since the publication of the papers *Toward an a-temporal interpretation of quantum potential* and *Toward a new interpretation of subatomic particles and their motion inside a-temporal physical space* in “Frontier Perspectives”, the authors of this book have suggested that the stage on which natural phenomena take place is an a-temporal physical space, that universe is an a-temporal phenomenon, and the Planck-length quanta of space are its elementary constituents]11, 12].

Whilst experimentation and theory-building continues, it is already safe to say that truly empty space is relegated to history. The reality recognized at the frontiers of physics is a cosmic plenum filled with universal forces and virtual particles. The observable and measurable world of particles and particle interactions is a subset of this plenum.

The floor of the universe is deeper than it was believed until recently. Underlying the manifest three-dimensional world of particles, forces and interactions there is a world that does not contain energy and matter in the known form, nor does it include space and time in the accepted sense.

2.2. About the Quantum Gravity Domain and Unifying Theories: Strings, Loops ... and the E8 Theory

General relativity and quantum theory have profoundly changed our view of the world. Furthermore, both theories have been verified to extraordinary accuracy in the last several decades. However, these theories appear as mutually incompatible and leave an image of the universe that is somewhat incomplete and fragmented.

Quantum theory assumes a fixed, background space-time geometry. Nevertheless, at the quantum realm, space-time geometry is not fixed, but it is a dynamical variable fluctuating and without a definite value. Quantum theory is very accurate, efficient and successful as far as the quantum phenomena do not interact with space-time, which plays the role of a background entity. Nevertheless, it breaks down when space-time is dynamical and interacts with the quantum phenomena. One needs therefore to construct a unitary theory describing the so-called quantum gravity domain, namely the gravitational interaction at Planck's scale.

In this regard, we can say that currently the mainstream of physics has not got a completely convincing quantum theory of gravity. There are instead some more or less developed approaches to this field, of which the most actively research fall into one of two broad classes: superstring theory and loop quantum gravity.

Superstring theory seeks to provide a unified quantum theory of all interactions on the basis of the idea that the elementary entities of the universe are not point particles, but one-dimensional extended objects called strings characterized by different vibrations inside a multi-dimensional background [13, 14]. Superstring theory arose from work on the strong interaction in the late 1960s and early 1970s when it was shown that all properties of a certain interesting model of the strong interaction (proposed by the Italian theoretical physicist Gabriele Veneziano) could be duplicated by a mathematical theory of a relativistic string. Though interesting, this idea did not really take off. In the mid-1970s, however, it was demonstrated that graviton-graviton[5] scattering amplitudes were the same as the amplitudes of a certain type of closed string. This fact led to the idea that superstring theory is a theory of all forces, not only of the strong interaction. Since the 1980s many researchers have contributed to this programme

---

[5] Standard quantum field theory postulated that gravitational field is transmitted by peculiar bosons called gravitons, that are therefore postulated as the quanta of the gravitational field.

of unification of all forces (but the father of this theory can be probably considered the British physicist, now at Cambridge University, Michael Greene).

Superstring theory is based on quantum mechanics and general relativity but does not consider the general-relativistic revolution regarding the background independence as fundamental. In order to obtain a quantum theory of gravitation, in this theory, the gravitational field is separated into two components: one component is regarded as a background while the other is treated as the quantum field. In this picture, superstring theory shows that a unified treatment of matter and forces is possible if we leave the hypothesis that the fundamental objects of nature have a pointlike structure. According to superstring theory, if we could examine the fundamental particles of the universe that physicists have identified – electrons, neutrinos, quarks, and so on – with even greater precision (a precision many orders of magnitude beyond our present technological capacity) we would find that each consists of a tiny, one-dimensional filament, the string, that vibrates and oscillates, namely a one-dimensional object with a finite spatial extent, which could not be described in terms of deeper constituents. It is important to underline that the simple replacement of point-particle material constituents with strings resolves the incompatibility between quantum mechanics and general relativity. Superstring theory therefore seems to unravel the central Gordian knot of contemporary theoretical physics. This is an important achievement, but it is only part of the reason superstring theory has generated such excitement.

In superstring theory strings represent the entire spectrum of particles through their vibrations. In this way, superstring theory promises a novel and attractive ontological unification. That is, unlike electrons, protons and neutrons – which together compose atoms – and unlike quarks and gluons – which together compose protons – strings would not be merely the smallest objects in the universe, the entities from which other types of matter are composed. Strings would not be merely constituents of electrons and protons in the same manner as these entities are constituents of atoms. Rather, strings would be all that there is: electrons, quarks, and so on would be simply different vibrational modes of a string. In the mid-1980s this idea was taken up by several researchers, who developed an unitary quantum theory of gravity in ten dimensions based on strings.

In Einstein's day, the strong and weak forces had not yet been discovered, but he found the existence of even two distinct forces —gravity and electromagnetism — deeply troubling. Einstein did not accept that nature is founded on such an extravagant design. This launched his 30-year voyage in search of the so-called "unified field theory" that he hoped would show that these two forces are really manifestations of one deeper underlying principle. This quixotic quest isolated

Einstein from the mainstream of physics, which, understandably, was far more excited about delving into the newly emerging framework of quantum mechanics. He wrote to a friend in the early 1940s, "I have become a lonely old chap who is mainly known because he doesn't wear socks and who is exhibited as a curiosity on special occasions". Einstein was simply ahead of his time. More than half a century later, his dream of a unified theory has become the Holy Grail of modern physics. And a sizeable part of the physics and mathematics community is becoming increasingly convinced that superstring theory may provide an important answer in this regard. From one principle — the idea that everything at its most microscopic level consists of combinations of vibrating strands — superstring theory provides a single explanatory framework capable of encompassing all forces and all matter. Superstring theory proclaims, for instance, that the observed particle properties — that is, the different masses and other properties of both the fundamental particles and the force particles associated with the four forces of nature (the strong and weak nuclear forces, electromagnetism and gravity) — are a reflection of the various ways in which a string can vibrate. Just as the strings on a violin or on a piano have resonant frequencies at which they prefer to vibrate — patterns that our ears sense as various musical notes and their higher harmonics — the same holds true for the loops of superstring theory. But rather than producing musical notes, each of the preferred mass and force charges are determined by the string's oscillatory pattern. The electron derives from a string vibrating one way, the up-quark derives from a string vibrating another way, and so on. Far from being a collection of chaotic experimental facts, particle properties in superstring theory are the manifestation of one and the same physical feature: the resonant patterns of vibration — the music, so to speak — of fundamental loops of string. The same idea applies to the forces of nature as well. Force particles are also associated with particular patterns of string vibration and hence everything, all matter and all forces, is unified under the same rubric of microscopic string oscillations — the "notes" that strings can play. In synthesis, superstring theory provides a framework with the capacity to explain every fundamental feature upon which the universe is constructed with the same underlying idea. For this reason superstring theory is sometimes described as a possible candidate of the so-called "theory of everything" or the "ultimate" or "final" theory. These relevant descriptive terms are meant to signify the deepest possible theory of physics — a theory that underlies all others, one that does not require or even allow for a deeper explanatory base.

On the other hand, contrary to superstring theory, in order to provide a quantum treatment of gravitation, loop quantum gravity directly faces the problem that, as expected by general relativity, indeed there is no background space in nature and builds a quantum field theory in a form that does not require a background space [15-19]. Loop quantum gravity consists thus in a direct application of quantum mechanics to general relativity: firstly, the gravitational field is identified with the metric structure of space-time and, then, this identification is extended also in quantum regime. General ideas on how to develop a quantum theory of gravitation that does not require a background space were put forward in the 1950s and the 1960s. Charles Misner, a theoretical physicist now at the University of Maryland, suggested that calculations on quantum gravity could be performed by summing over all possible space-times, in analogy to Feynman's version of quantum field theory[6]. John Archibald Wheeler suggested that space-time has a foam-like structure at very small scales and, together with Bryce DeWitt of Texas University, introduced the idea of a "wave function over geometries", a function that expresses the probability of having one space-time geometry rather than another. This wave function over geometries obeys a very complicated equation that is now called the Wheeler-DeWitt equation, which is the fundamental equation describing the quantum gravity and cosmology domain and which practically is a sort of Schrödinger equation for the gravitational field. These brilliant ideas became concrete at the end of the 1980s, when a well-defined mathematical theory describing quantum space-time characterized by a background independence, namely loop quantum gravity, began to form. In this regard, it is important to mention the contributions, above all, of Carlo Rovelli, Italian physicist now at the University of Marseille, of the Indian physicist Ashtekar, of the English physicists Penrose and Isham, of the Scottish physicist Pullin, of the Polish physicist Lewandowski and of the American physicists Smolin and Baez.

The most appealing aspect of loop quantum gravity is that it predicts that space is not continuous, is not indefinitely divisible, but that it has a granular structure, given by a network of intersecting nodes, "loops" having the size of Planck's length. These loops represent the quantum excitations of the gravitational field namely the elementary quanta of space. Loop quantum gravity is therefore the first theory of fundamental physics that provides the evidence that, at Planck's scale, space is quantized, namely that a measurement of area or volume of a region of space can give only a discrete spectrum of values and that we cannot observe an area or a volume which is lower than Planck's scale. Moreover, just like quantum mechanics allows us to determine precisely, for example, the energy levels of the hydrogen atom, in analogous way loop quantum

[6] Feynman's version of quantum field theory is characterized by the famous Feynman's diagrams in which each particular graphic element corresponds to a specific physical and mathematical object.

gravity allows us to compute explicitly the spectrum of values that area and volume of space can assume, providing therefore detailed empirical predictions of physics of the Planck scale. In particular, in this theory, quantum numbers expressed by integer or half-integer numbers are associated with the separation surface between each couple of adjacent nodes. Since these quantum numbers and their algebra looked like spin angular momentum numbers of elementary particles, Penrose called them "spin networks".

In loop quantum gravity, the spin-networks picture of space is mathematically precise and physically compelling: nodes of spin networks represent the elementary grains of space, and their volume is given by a quantum number that is associated with the node in units of the elementary Planck volume. Two nodes are adjacent if there is a link between the two, in which case they are separated by an elementary surface the area of which is determined by the quantum number associated with that link. Link quantum numbers are integer or half-integer and the area of the elementary surface is just linked to these quantum numbers as well as to the Planck volume.

It is important to underline that in loop quantum gravity it makes no sense to say that loops are collocated in space, because here there is no fixed background space. Loops are themselves space because are the quantum excitations, the elementary quanta of the gravitational field. Therefore, only the relative position of a loop with respect to another loop has sense. The picture of physical world that is provided by loop quantum gravity is practically represented by a network of intersecting loops, of intersecting nodes (and there is no fixed background in which they are collocated).

Moreover, if in quantum mechanics we have the superposition principle (according to which the state of a physical system can be given by the superposition of more terms corresponding to perceptively and macroscopically distinguishable situations), in analogous way in loop quantum gravity the general state of physical space is given by a quantum superposition of different terms corresponding to different networks of loops (even if at the Planck scale the things are more complicated). In synthesis, we can say that in loop quantum gravity a physical region of space is describe by a quantum superposition of spin-network states, and the dynamics in the region are governed by a well-defined Wheeler-DeWitt equation.

As regards the view of time (as we have already said in chapter 1.1) at Planck scale a universal time is not defined. In loop quantum gravity there is no independent temporal dimension along which dynamics happens, there is not a well-defined concept of time and thus time must be re-introduced as a reading on a determinate physical clock. Although in loop quantum gravity an idealized, universal time does not exist, several physicists managed to develop interesting models of spin-network states, known as "spinfoam models" that permit to visualize the evolution of physical space. In these models, space-time is seen as a geometrical sequence of spaces, as a history of spaces and therefore is described by means of a history of spin-networks. The term spinfoam is used to indicate just the history of spin networks. It has a simple geometrical interpretation. It is constituted by points, lines and surfaces: the history of a point is a line and the history of a line (namely of a link between two elementary adjacent grains) is a surface. In synthesis, we can therefore say that loop quantum gravity replaces the Newtonian concept of space-time intended as absolute background with a history of spin networks of Planck scale, called spinfoam.

Moreover, building on the insights of loop quantum gravity, in 2007 the itinerant American physicist Anthony Garrett Lisi – who balances his time research in theoretical physics and surfing – proposed a new unified theory which met a significant resonance. This unified field theory, called E8 theory, allows us to go beyond the Standard Model, by considering that all forces and matter are described as the twisting of a single geometric object [20].

E8 theory follows the methods of quantum field theory and loop quantum gravity. One enticing possibility is that the gravitational and cosmological constants run from large values at an ultraviolet fixed point to the tiny values one encounters at low energies. At the foundational level, a quantum description of the standard model in E8 theory may be compatible with a spin foam description in terms of braided ribbon networks. This theory represents a comprehensive unification program, describing all fields of the standard model and gravity as parts of a uniquely beautiful mathematical structure. The mathematical objects of this approach describe how the manifold postulated by the E8 theory twists and turns over space-time, reproducing all known fields and dynamics through pure geometry. Although some aspects of this theory are not yet completely understood, the match to the standard model and gravity seems very good. This theory can unveil some of the beautiful structures in play at the deepest levels of nature, and make predictions for new particles that the Large Hadron Collider might find. E8 theory seems therefore to be an important step on our centuries-long quest for unity.

In E8 theory, the geometric objects associated with elementary particles are smooth shapes, existing outside our space yet connected to it. We cannot see these shapes directly, but we see their effects. The main geometric idea underlying the Standard Model is that every point in our space-time has shapes attached to it, called fibers, each corresponding to a different kind of particle. The universe can be envisioned as a Chia Pet (a terra-cotta figurine covered with sprouts). The surface of the figurine is analogous to space-time, and the sprouts are fibers. The entire geometric object – a space-time and fibers together – is called a fiber bundle. The fibers are not in our space, but over it; they may be thought of as different, internal spaces attached to each point of our space-time, with shapes corresponding to particles' properties.

This idea, introduced in 1918 by mathematician Hermann Weyl, is now a well-established principle of physics. Distinct from the speculated undulating spatial dimensions of string theory, these internal space fibers are of fixed shape. Their dynamics arise from how they are attached to four-dimensional space-time.

The electric and magnetic fields existing everywhere in our space are the result of fibers with the simplest shape: the circle. The fiber bundle of electromagnetism consists of circles attached to every point of space-time. Crucially, each circle can rotate a little with respect to its space-time neighbours. A fiber bundle is characterized by a connection field which describes how neighbouring fibers are related by symmetry rotations. The electric and magnetic force fields filling space-time correspond to the curvature of this fiber bundle – geometrically, the electric and magnetic fields are how the circular fibers twist over time and space. An electromagnetic wave is the undulation of circles over space-time. One quantum of an electromagnetic wave – a photon – is a propagating particle of light.

In E8 theory, each kind of elementary particle corresponds to a peculiar fiber over space-time. The fibers of electrically charged particles, such as electrons, wrap around the circular fibers of electromagnetism like threads around a screw. How fast a particle's fiber twists around the circle is equal to its electric charge, determining how the particle responds to the force of electromagnetism.

Moreover, as regards the treatment of gravity in this geometric approach, we can imagine that as we move to different points in space-time, there are different sets of rulers and clocks, related to our original frame by a rotation. This rotation can be an ordinary rotation in space or a rotation of space into time. How the frame rotates from point to point is determined by a physical structure of space, which is called the gravitational spin connection. We feel the force of gravity because the gravitational spin connection field is rotating our frame as we move through time, attempting to steer us toward Earth's centre.

Just as particles have different kinds of charge describing how they interact with the interactions of the Standard Model (namely electromagnetism, weak force and strong force), they have a type of charge describing how they behave within space. Consider what happens if we rotate a ruler in space by 360 degrees: it returns to its original state. This ruler — and the gravitational frame field — has spatial spin charge of +1 or –1. But if we rotate a fermion, such as an electron, in space by 360 degrees, it does not return to the same state it started in. To return it to its original state, we have to rotate it by 720 degrees. The fermion has spin charge of ±1/2. Spin charge plays a role in gravity because gravity, through the frame and spin connection, is related to the geometry of space-time. A particle's spatial spin charge is its internal angular momentum, and its temporal spin charge is related to its motion through space. Fermions whose spatial spin and motion align make a right-handed corkscrew as they travel through space. Fermions with opposite motion and spatial spin are left-handed. Spin charge also has an unexpected relevance to the weak nuclear force. Only left-handed particles and right-handed antiparticles have weak charge and interact with the weak force. The fact that the weak force is sensitive to spin charge suggests that gravity and the other forces, though outwardly dissimilar, in fact have a deep relationship.

In this way, in E8 theory the physical universe could emerge naturally from a mathematical structure without peer. This theory explains in a unitary picture what Higgs bosons are, how gravity and the other forces emerge from symmetry-breaking, why fermions exist with the spins and charges they have, and why all these particles interact as they do [20, 21].

2.3. Modelling Physical Reality at the Fundamental Level: Being Versus Becoming and Atomism Versus Monism

In the quest to unify all of physics, the combination of Leibniz's principle of sufficient reason (which states that "no fact can be real or existing and no statement true without a sufficient reason for its being so and not otherwise") plus the dynamical perspective as the ultimate explanation of processes has in great part motivated the particular kind of unification being sought, i.e., the search for a TOE, quantum gravity theory or similar. Almost all attempts to unify relativity and quantum physics which can be found in literature opt for becoming (dynamism) as fundamental in some form or another. Such theories may deviate from the norm by employing radical new fundamental dynamical entities (such as strings, loops, or fibers inside Garrett-Lisi's E8 theory, etc.), but the game is always dynamical, broadly construed. Almost all fundamental theories of physics imply a fundamentality of becoming versus being as key of modelling and describing physical reality. From the perspective of these theories, the idea of a becoming universe implies a view of space and time that not only distinguishes the future events from the past ones intrinsically, but also accounts for the continual passage of the fleeting present, from a non-existing future into the unalterable past. Such a theory of space and time, which would account for the gradual coming-into-being of the non-existent future events – or a continual accumulation of the unalterable past ones – giving rise to a truly becoming universe, may be also defined as a Heraclitean theory of space-time, in order to be distinct from a Parmenidean one, devoid of any such explicit dictate to becoming.

Dynamism indeed encompasses three claims:

A) the world, just as appearances and the experience of time suggest, evolves or changes in time in some objective fashion,
B) the best explanation for A will be some dynamical law that "governs" the evolution of the system in question, and
C) the fundamental entities in a TOE will themselves be dynamical entities evolving in some abstract space. However, in spite of the presumption of dynamism, those who want fundamental explanation in physics to be dynamical and those who want a world that evolves in time in some objective fashion, must face all the matters concerning the timeless perspectives in physics analysed in chapter 1, as well as other well-known problems such as:
    1. the possible blockworld implications of relativity (both special and general) and
    2. quantum gravity, the quantization of a generally covariant classical theory leading to the famous problem of time (also called "frozen time").

Also as regards special relativity, given certain widely held innocuous assumptions, the Minkoswki formulation seems to be in contrast with the view of a fundamental dynamism. To use the words of the American physicist Robert Geroch of the university of Chicago [22]: "There is no dynamics within space-time itself: nothing ever moves therein; nothing happens; nothing changes. In particular, one does not think of particles as moving through space-time, or as following along their world-lines. Rather, particles are just in space-time, once and for all, and the world-line represents, all at once, the complete life history of the particle."

A way out to this tension between blockworld and frozen time on the one hand and dynamism on the other could be of rejecting the former whether conceptually or formally and attempting to construct a fundamental theory that has something definitively dynamical at bottom. The idea subtending this programme is to make time or change fundamental in some way, as opposed to merely emergent as in the case of string theory or an illusion as in the case of the Wheeler-DeWitt equation. Smolin, for example, suggests a radically "neo-Heraclitean" solution wherein change and becoming are fundamental in that axiomatic dynamical laws, the values of constants that figure in those laws and configuration space itself evolve in time or meta-time [23]. Though he does not necessarily frame it this way, Smolin is advocating for something like a process-based physics where change or flux itself is fundamental. In doing so, Smolin joins Bohm and Hiley who have been advocating such an approach for many decades.

In the context of the fundamental physical theories describing the ultimate quantum foams subtending the observable forms of matter, energy and space-time, strictly connected with the debate between being and becoming is the debate between atomism and monism. In this regard, despite all the tension that quantum theory has created for atomism as originally conceived, most physicists still assume there is something fundamentally entity like at bottom. But on the process view, potentia, activity, flux or change itself is fundamental, not entities/things changing in time such as particles or strings. In this monistic physics, all talk of such dynamical entities would emerge from, and be derived from, the more fundamental flux together with, and inseparably from, space-time in a background independent fashion. For example, as we have already mentioned in chapter 1.8, Bohm and Hiley developed a monistic model wherein "the whole is prior to its parts, and thus views the cosmos as fundamental, with metaphysical explanation dangling downward from the One" [24].

Hiley's implicate order extends Bohm's interpretation of quantum mechanics to the relativistic regime and unifies space-time geometry and material processes. In fact, Hiley's approach is motivated by the following starting idea: he does not want things happening in a background space-time but wants to "start from something more primitive from which both geometry and material process unfold together". That which Hiley considers "more primitive" is elementary process. Hiley calls the fundamental process/potentia the "holomovement" and it has two intertwined aspects: the "implicate order" (characterized algebraically) and the "explicate (or manifest) order" which contains all the physics derived from it, such as space-time geometry. The holomovement is thus the whole ground form of existence, which contains orders that are both implicate and explicate, wherein the latter expresses aspects of the former [25].

From the perspective of the implicate order, rather than point particles being evolved in time aided by instantaneous updating by the quantum potential or pilot wave, the fundamental evolution is one of processes that give rise to explicate structures extended in space and time. In short, particles and pilot waves are not fundamental but are emergent from the implicate order. In the picture of Hiley's implicate order, there is no unique "now" successively coming into existence. Instead there are infinite time-like foliations of the Minkowski space-time manifold, each representing a unique global "now" at various values of its foliating time, and a particular spatial hypersurface in a given foliation A (a "now" for a given observer A) contains events on many different spatial hypersurfaces in foliation B (different "nows" for observer B). Events which are simultaneous for observer A are not simultaneous for observer B and this fact implies a "relativity of simultaneity" and negates an objective passage of time. That is to say, there is no objective (frame independent) distinction in space-time between past, present and future events respectively and therefore no objective distinction to be had about the occurrence or non-occurrence of events.

As regards the problems connected with time, in Hiley's view the idea is that the holomovement can explicate either a small region or a large region of space-time (to include the future) "at once", but never the entire universe. The extent of the explicate domain depends on the properties of the holomovement in each particular case and the process is apparently stochastic. Hiley's approach implies that what happens in the future cannot be made to rewrite the past, but that the future possibilities can influence the unfolding of the present. What is less clear is whether these moments pass in and out of existence or always stay in existence once explicated. Hiley's view suggests that each individual shadow manifold is constantly changing in its own time (evolving "now") such that the past is consistent with the present and the future is understood probabilistically.

On the other hand, the American physicists of Elizabethtown College Michael Silberstein, W. M. Stuckey and Timothy McDevitt proposed recently a monistic approach of unification of relativity and quantum theory, called the Relational Blockworld, in which the unity of space, time and matter is postulated at the fundamental level and is used to recover dynamical or process-like classical physics only statistically [26-28]. Silberstein's, Stuckey's and McDevitt's approach, unlike Hiley's approach, emphasizes being over becoming and has something closer to geometry at bottom (in the form of a discrete graphical structure). In this approach, the general relativity domain emerges, unlike Hiley's account, in a picture that is fundamentally adynamical and acausal, in contrast also to other fundamental theories attempting to quantize gravity (string theory, loop quantum gravity, etc.). According to the relational blockworld approach, quantum physics is the continuous approximation of a more fundamental, discrete graph theory: this model provides a wave-function-epistemic account of quantum mechanics with a time-symmetric explanation of interference inside an acausal picture. Quantum physics is simply providing a distribution function for graphical relations responsible for the experimental equipment and process from initiation to termination. So, EPR-correlations and the like evidence superluminal information exchange (quantum non-locality), in the Relational Blockworld approach, are actually evidence of the deeper graphical unity of space-time-matter responsible for the experimental set up and process, to include outcomes.

While the monistic character of Hiley's process-oriented approach is housed in the implicate order, the monistic character of the relational blockworld approach is housed in a space-time-matter arena which underwrites the space-time + matter classical world of our observations. Both the implicate order and the Relational Blockworld are approaches that want to explain such observed quantum phenomena from a more fundamental theory underneath quantum theory itself, although from the point of view of the dichotomy of becoming versus being they are quite opposing fundamental theories. Both approaches want to derive general relativity and quantum theory from something more fundamental in a background independent fashion such that the explanation for quantum entanglement and EPR correlations, rather than creating tensions with space-time and relativity, requires neither non-locality nor non-separability in space-time. Rather, such quantum effects are explained at the more fundamental level whether graphical (in the Relational Blockworld approach) or algebraic (in Hiley's implicate order approach).

According to the Relational Blockworld approach, quantum theory and relativity theory both tell us that every facet of dynamism is false, that dynamism emerges just as a historical contingency based on the fact that all physics must start with experience. Perhaps this model offers a fourth possibility regarding the nature of time, i.e., time as part of a fundamental (pre-geometric) regime wherein the notions of space, time and matter are co-defined and co-determining. In this picture, time, space and matter as stand-alone concepts are not fundamental, emergent or illusions; this model, being fundamentally relational, implies that there are no perspectives "external to the universe". Here, the universe comes with many physically significant modes of temporal passage and change such as proper time, cosmic time, etc… Certainly these constitute objective notions of becoming (objectively dynamical flow) even if they are mere patterns in a block universe. Therefore, the Relational Blockworld approach does not negate change and becoming, it merely internalizes and relativizes them. In the Relational Blockworld universe, everything, past, present and future is just there "at once" (including conscious experiences throughout the block) and brains are just worldtubes like everything else. It suggests that conscious temporal experience is fundamental as well, so instead of space-time-matter at bottom one has the super-monistic space-time-matter experience at bottom.

Another interesting physical model which implies the fundamentality of being versus becoming is Ruth Kastner's possibilist transactional interpretation of quantum mechanics. The transactional interpretation of quantum mechanics was first proposed by the American physicist John G. Cramer in a series of papers in the 1980s [29-32]. According to Cramer's approach, a quantum process is a time-symmetric process, in which we have an "absorber" which generates confirmation waves in response to an emitted offer wave: a system emits a field in the form of half-retarded, half-advanced solutions to the wave equation, and the response of the absorber combines with that primary field to create a process that transfers energy from the emitter to the absorber. Cramer's theory leads to a definition of a vacuum which gives form to the anticipated waves; these inform a particle about the global situation in its space-time region and in this way one can reproduce the quantum behaviours, and in particular explain EPR-type experiments in terms of information – the transaction events – coming from the future owed to the relativistic nature of the vacuum.

Ruth Kastner of the University of Maryland extended Cramer's approach into a more general and fundamental theory, the possibilist transactional interpretation, which can explain the nature of the process leading an actualized transaction [33]. In Kastner's approach, space-time is not a pre-existing substance, a structured container for events, but rather unfolds as an emergent manifold from a more fundamental collective structure, a "pre-spacetime" characterized by "transactional processes" (constituted by emission and absorption of quanta) involving de Broglie waves. In the picture proposed by Kastner, space-time intervals are measuring systems of realized transactions resulting in transfers of energy from an emitter to an observer, which must be considered as pre-space-time objects at the micro-level. In Kastner's view, the transactions are processes that somewhat transcend the space-time structure, in other words are the expression of the non-local feature of the quantum processes. In this model, space-time is no more – and no less – than the set of actualized transactions. Thus, actualizations of transactions, and thus of emissions and absorptions of quanta, give rise to the set of related events comprising the space-time arena. The actualization of a transaction is an a-spatiotemporal process; it is the coming into being of an event (actually two linked events, the emission and the absorption).

In an actualized transaction, the emission defines the past and the absorption defines the future. That is, past and future supervene on actualized transactions; there is no "space-time" without actualized transactions. The apparent four-dimensional space-time universe we perceive is not something "already there"; rather, it crystallizes from an indeterminate (but real) pre-space-time of dynamical possibility. Thus, space-time "grows" but not in the usual "growing universe" sense wherein an advancing "now" proceeds from present to future; rather, events arise from a set of dimensions defining the quantum domain outside space-time. In fact, it is the past that "grows" and is extruded from the present; in the pre-spacetime there is no actualized future. If, as a contingent fact of our universe, all processes associated with emission of quanta are provided with at least one absorber, then all emitters must ultimately exist within space-time since they will participate in an actualized transaction. However, absorbers corresponding to unactualized transactions – that is, transactions that "failed" – do not exist within space-time (unless the same absorber participates in an actualized transaction from some other quanta emission).

Kastner's possibilist transactional interpretation of quantum mechanics implies that only the "now" is the fundamental empirical realm of physical processes and that the changing properties of physical systems are associated with electromagnetic signals that transfer energy from what we are observing to our sense organs (by way of actualized transactions). The "now" is defined by a spatial coordinate (or, in a relational view of space-time, the object(s) with which we are currently in direct contact) and any light signals that have reached our eyes from other objects. Moreover, the empirical realm indeed cannot be "objective" in absolute terms in the sense that it is defined in terms of appearances which can only be relative to a given observer. This means that, strictly speaking, everyone has a different "empirical realm." However, we can corroborate our experiences and arrive at a consistent intersubjective consensus about a "larger" world of appearance beyond our individual empirical realms. All of these corroborations are associated with electromagnetic waves. Thus, what is referred to as "the empirical realm" of physical processes can be no more than that corroborated collection of individual empirical "nows".

According to the possibilist transactional interpretation, what is fundamental in the temporal sense is just the now, which changes. The now, and thus being is the ultimate, primary reality of the universe, and the time index is just a way of recording the changes of the now. In other words, one deals here with a sort of "growing universe" theory of time; the past grows and continues to become actualized as it falls away from the present. This approach implies a fundamentality of being versus becoming in the sense that one deals with a blockworld of "nows" originating actualized transactions. The future is a realm of physical possibilities. The present does not "advance" into the future. Rather, the future is a set of possibilities that becomes woven into the created past through the action of now. The origin of time is contained in matter. According to the possibilist transactional interpretation, if there were no matter, there would be no spacetime. It is matter that creates the separation between temporal and spatial axes through transactions.

2.4. THE POSTULATES OF THE THEORY OF EVERYTHING

One of the perennial problems in natural philosophy is the problem of change; namely, how is change possible in natural processes? Over the centuries, this problem has fostered two diametrically opposing views of time and becoming. While these two views tend to agree that time presupposes change, and that genuine change requires becoming, one of them actually denies the reality of change and time, by rejecting becoming as a "stubbornly persistent illusion". The other view, by contrast, accepts the reality of change and time, by embracing becoming as a bona fide attribute of the world. Since the days of Aristotle, physics has been rather successful in explaining how the changes occur in the universe, but seem to remain oblivious to the deeper question of why do they occur at all. On the basis of the current research mentioned and analysed in chapter 1, a new key of reading is opened as regards the views of time and becoming: time exists only as an emergent mathematical quantity measuring the numerical order of material changes inside an universe that, at a fundamental level, is timeless, but change is not an illusion. The fundamental reality is represented by the processes of creation and annihilation of quanta from a fundamental timeless three-dimensional quantum vacuum. In this picture, at a fundamental level, just like in the Relational Blockworld approach or in Kastner's transactional interpretation, being reigns over becoming, in the sense that RS processes are the only primary physical reality. At the explicate order of the macroscopic domain time is an emerging quantity and thus the becoming characterizing the events can be considered as an emergent reality. But the emphasizing of being over becoming here gets a different significance with respect to Silberstein's, Stuckey's and McDevitt's Relational Blockworld approach, in the sense that in the picture of the universe proposed by the authors of this book there is not an incompatibility between being and becoming: here we deal rather with a peculiar unitary complementarity in which the becoming of the events, of motion and evolution derives from the primary physical reality represented by the RS processes of creation-annihilation of quanta in the timeless three-dimensional quantum vacuum. In the view proposed by the authors of this book it is so possible to clarify in what sense – as established by Kastner's transactional approach – the origin of time is contained in matter. Matter creates the separation between temporal and spatial axes through transactions in the sense that time emerges as a numerical order of material changes from the primary physical reality represented by the RS processes of creation-annihilation of quanta in the timeless three-dimensional quantum vacuum. As a consequence, one can say that, in this picture, being and becoming represent indeed two different aspects of the same coin.

We consider that the fundamental arena of the universe is a three-dimensional space originating in a three-dimensional quantum vacuum where fundamental time is a numerical order of change and emergent time is duration of material change running in space. Time is not a fourth physical dimension of universal space. Such an approach can be considered as a possible standpoint in order to bridge general relativity and quantum electrodynamics.

In this chapter and the next ones, on the light of the current physics and of some recent research of the authors of this book, we want to see how one can introduce some interesting perspectives towards a unitary timeless theory of everything which embraces these foundations.

Taking account of the relevant results of fundamental physical theories such as superstring theory and of loop quantum gravity, of a realistic unified vacuum subtending the observable forms of matter and energy as well as of a timeless three-dimensional background as fundamental arena of the universe, according to the authors it is legitimate to assume that the fundamental reality is a "space-matter" arena, composed by elementary packets of energy at the Planck scale, which emerges from a timeless three-dimensional quantum vacuum characterized by RS processes and by changes of its energy density. As a consequence of the timeless feature of the fundamental level of physical processes represented by the three-dimensional quantum vacuum, matter and space are composed by the same type of elementary constituents – quanta of space having the size of Planck's length – which must be intended as timeless constituents. Quanta of space, as elementary packets of energy, have not been created and cannot be destroyed, but can only transform themselves from a "form" to another. Each quantum of space contains a specific quantum information which corresponds to a specific RS process of creation or annihilation from the three-dimensional quantum vacuum. Universe is a system in dynamic equilibrium which is characterized, because of the information associated with the three-dimensional quantum vacuum, by the cyclic transformation "space-matter-space-matter-...".

The model developed by the authors of this book – which can be called as "theory of everything in the timeless space" – opens the perspective that both timeless empty space and matter are composed by quanta of space of the Planck length, endowed with different frequencies of vibration and, consequently of different "states" of energy [34-38, 11,12]. The information contained in each quantum of space can be expressed in the form of particular vibrations and, as a consequence, of different states of energy. The manner in which a frequency of vibration is associated with a specific energy is that it provides the physical possibility for a transfer of that specific energy from one quantum to another and consequently to determine motions and produce potentially evolutionary phenomena. In this model, quanta of space constituting empty timeless space – far away from material objects – are vibrating with the so-called "basic frequency" $\nu_p = 0{,}19 \cdot 10^{44}$ per second, which coincides with the inverse of Planck's time and thus is the maximum frequency existing in nature. Quanta of space characterized by the basic frequency are associated with a maximum energy density of the three-dimensional quantum vacuum. They change their electrical charge from positive to negative in a Planck time ($5.39 \cdot 10^{-44}$ seconds), which is the lowest fundamental unit of time, namely of motion. Quanta of space constituting timeless empty space are complete into themselves: their existence does not depend on other physical entities. They have no radiation, no "dispersion of the energy": their energy is always the same and exactly the basic energy $E_{QS} = 1{,}26 \cdot 10^{10}$ Joules. This means that quanta of space building up timeless empty space have no entropy, in other words can be defined as the "non-entropy state" of energy. It is in virtue of these features that they are not perceivable to our senses and space appears to us just empty, without changes.

The assumption that the basic frequency corresponding to the inverse of Planck's time play a crucial role in the timeless space can be considered legitimate also in virtue of the fact that, according to current research, this peculiar frequency acts as an intermediate reality between two fundamental theories of physics as special relativity and Hamiltonian mechanics. For example, the English physicist Christian of Oxford University since 2007 has provided a generalized theory of the inertial structure in which proper time is causally compelled to be tensed within both space-time and dynamics in a picture where the inverse of Planck time is introduced as the conjunction between special relativity and Hamiltonian mechanics and lengths and durations are bounded by their respective Planck scale values. Christian's approach abhors any form of preferred structure, and yet captures the transience of now along timelike worldlines by causally necessitating a genuinely becoming universe [39].

In the view of the authors of this book, if quanta of space corresponding to regions where the energy density of the three-dimensional quantum vacuum is maximum, instead quanta of space constituting matter (which correspond to regions characterized by changes of the quantum vacuum energy density and by opportune RS processes of creation) vibrate with appropriate different frequencies (lower than the basic one) and are endowed with entropy (they can be thus defined as the "entropy-state" of energy). For this reason, matter is perceivable to our senses and is subjected to changes (for example in the position and in the speed). First law of thermodynamics rules quanta of space constituting timeless empty space (and thus regions where the energy density of the three-dimensional quantum is maximum), second law of thermodynamics rules quanta of space which build up matter. There are finally also other physical entities endowed with entropy, namely electromagnetic waves, that can be interpreted as entities which derive from vibrations of quanta of space with the frequencies of the electromagnetic spectrum, that are propagating in the timeless space with the light speed.

In synthesis, according to the research of the authors of this book, a "theory of everything in the timeless space" can be suggested by starting from the following postulates [35-38, 40-42, 11, 12]:

1. Energy is composed by basic packets of energy which have the Planck size (here called "quanta of space"). Quanta of space constituting physical space have no entropy.
2. Quanta of space form physical space and matter. The "arena" (stage) of the universe is the gravitational space, namely a "space-matter" arena. Physical reality is composed by space and matter. Quanta of space contain a quantum information which is linked with a wave-function depending on its position in physical space, a quantum number which indicates its rotation-orientation (with respect to an arbitrary axis) and a frequency of vibration. Quanta of space vibrating with the basic frequency constitute "empty" physical space, are devoid of mass and are the "non-entropy state" of energy and therefore are not perceivable by our senses. Quanta of space vibrating with the frequencies characteristic of material particles constitute matter and are the "entropy state" of energy (and therefore are perceivable to our senses). Each subatomic particle corresponds to a region characterized by an opportune quantum vacuum energy density and appears in virtue of opportune RS processes of creation from the three-dimensional quantum vacuum. The origin of subatomic particles from the processes of the fundamental three-dimensional quantum vacuum implies that each subatomic particle can be seen as the result of the interaction of energy in the "entropy state" with one or more quanta of space, caused by the vibration of these quanta of space at appropriate frequencies. In particular, particles devoid of internal

structure, such as quarks, leptons[7] and intermediate bosons[8] are the result of the interaction of energy in the entropy state with one quantum of space; particles endowed with an internal structure, such as baryons and mesons, are given by the interaction of entropic energy with more quanta of space. Quanta of space vibrating at the frequencies of the electromagnetic spectrum generate electromagnetic waves (which propagate through physical space at the speed of light).

3. Time is numerical order of material change that runs in space. Space itself, at a fundamental level, is timeless. The fundamental arena of the universe is a three-dimensional space which originates from a three-dimensional quantum vacuum – characterized by RS processes of creation and annihilation of quanta and by changes of the quantum vacuum energy density – where fundamental time is a numerical order of change and emergent time is duration of material change running in space. In the universe space and matter are in a permanent dynamic equilibrium. There was no beginning of the universe and there will be no end.
4. The fundamental interactions and physical fields represent special states of space, the different ambient situations existing in space in presence of certain material particles. All interactions produce, determine modifications in the properties of space. In particular, gravity has the effect to produce modifications in the geometrical properties (i.e., in the curvature) of space. The other three interactions (electromagnetic, weak and strong) determine modifications in the vibrations of quanta of space. They can change the frequencies of quanta of space from the "basic frequency" (characterizing empty space) to other appropriate frequencies less than the basic one (which can be the frequencies characteristic of material particles and/or of electromagnetic waves), or can change the frequencies characteristic of some material particles in the frequencies characteristic of other material particles (and/or of electromagnetic waves). Each interaction is characterized by its own strength parameter, which indicates the intensity of the modifications induced in the region of space under study, and its own particular range, i.e., the range in which the modifications of space determined by the interaction happen.

[7] Leptons are particles such as the electrons. Together with quarks, they constitute a peculiar set of fundamental particles of physics known as fermions, that satisfy a specific mathematical apparatus known as Fermi-Dirac statistics and have a half-integer spin.

[8] Bosons are the particles which transmit the different interactions, satisfy the mathematical apparatus known as statistics of Bose-Einstein and are endowed with a integer spin.

Gravitational force is an immediate physical phenomenon transmitted directly by the quantum vacuum energy density; material objects move in the direction in which the quantum vacuum energy density is decreasing. The other forces are mediated each by the exchange of a particular boson: the photon for the electromagnetic interaction, the intermediate bosons $W^{\pm}$ and Z^{0} for the weak interaction and the gluons for the strong interaction.

5. The description of a force is not altered by any modification of the length scales of rulers and of temporal scales of clocks utilized as measurement instruments (gauge invariance principle). Each interaction satisfies its peculiar gauge symmetry. (For example, in the case of the electromagnetic interaction, the quantum-mechanical description of experiments on charged particles is invariant under local phase transformations on the particle wave function, if one introduces a long-range field coupled to the charge – the electromagnetic field – and one makes simultaneously a suitable local gauge transformation on the electromagnetic potential.).

2.5. The Interpretation of Subatomic Particles in the Timeless Space

In the theory of everything suggested by the authors of this book, the quantum information contained in the quanta of space, and in particular their frequencies of vibration can be considered as the primary elements, the fundamental "forces" that determine the features of the entities present in the universe. This view opens the perspective that a quantum of space, if vibrates with the basic frequency, is a constituent of the timeless empty space, devoid of matter; if it vibrates with a frequency characteristic of a subatomic particle, it is a material quantum of space, endowed with mass and thus of entropic energy; and, finally, if it vibrates with the frequency of the electromagnetic spectrum, it satisfies the equation of electromagnetic waves, namely is a constituent of electromagnetic radiations.

In synthesis, the model developed by the authors suggests that the frequencies of vibration of the quanta of space are the fundamental physical variables, in other words the fundamental source of the information that determine the properties of the objects we observe and perceive. We can say that we practically live in a sort of "sea of frequencies": our perception of the different entities is linked with the different frequencies of vibration of the quanta of space that constitute them.

All material objects are different portions, aggregates of quanta of space with their own characteristic frequencies of vibration. All material particles are composed by quanta of space that vibrate at appropriate frequencies and that, in virtue of these vibrations, become seat of a discrete quantity of energy in the "entropy-state". While quanta of space vibrating at their basic frequency have got energy in the "non-entropy state" (and constitute therefore "empty" timeless space corresponding to regions where the quantum vacuum energy density is maximum), quanta of space vibrating at appropriate frequencies (lower than the basic one) assume energy in the "entropy state" and become thus material quanta (corresponding to regions characterized by changes of the quantum vacuum energy density and by opportune RS processes), endowed with mass and therefore perceivable by our senses.

In this model based on a timeless space, each subatomic particle is therefore interpreted as the result of the interaction of energy in the "entropy-state" with one or more quanta of space, determined by the vibrations of these quanta of space at appropriate frequencies. More precisely one can propose that particles devoid of internal structure, such as quarks, leptons and intermediate bosons, can be seen as the result of the interaction of energy in the "entropy-state" with one quantum of space (caused by the vibration of this quantum of space at a certain appropriate frequency); instead, particles endowed with an internal structure such as baryons (constituted by three quarks) and mesons (constituted by a quark-antiquark pair) are given by the interaction of entropic energy with more quanta of space (caused by the vibration of these quanta of space at certain appropriate frequencies). For example, electron can be seen as the result of the interaction of energy in the "entropy-state" with one peculiar quantum of space (characterized by the peculiar change of the quantum vacuum energy density and by the RS process corresponding to the appearance of the electron), interaction determined by the vibration of this quantum of space at an appropriate frequency.

An interesting perspective introduced by the theory of everything in the timeless space is therefore the following: when particles move, it is not an actual physical thing that changes position in space, but the vibratory state that creates the particles (and that emerge from a corresponding change of the quantum vacuum energy density and a corresponding RS process of the same three-dimensional quantum vacuum) moves across a range of space quanta, successively re-creating the particle at each new location. The motion of the vibratory state corresponding to a given subatomic particle can be seen as a consequence of the occurring of a sequence of fundamental RS processes of creation and annihilation. Here, one can remark an interesting analogy with the proposal of another Italian physicist, Gian Piero Godone, who in 1994 posited a fourth law of motion, to add to Newton's three: "Any possible movement of any physical body is due to the constant motion of its basic elements, which are continually substituted by always new basic elements that arise in an orderly sequence from space and then to there return."

Another interesting perspective lies in the fact that the discrete quantity of entropic energy that a quantum of space assumes when it originates a material particle (in correspondence with fundamental RS processes of creation or annihilation) depends on the features of the region into examination, on the environmental situation characterizing that particular region of space (namely on the type of interaction, "potential" to which the region of space is subjected). In fact, one can say that it is the special state characterizing the timeless space (and which derive from the fundamental processes of the timeless three-dimensional quantum vacuum) to determine the appearance of a given particle in a given portion of space. The environmental situation (namely the physical field) characterizing a certain region of physical space can produce changes in the properties of that region. Here, the fundamental interactions and physical fields can be interpreted as special states, as special environmental situations characterizing the timeless space in presence of certain material particles, and can produce modifications in the properties of space. In particular, gravity has the effect to produce modifications in the geometrical properties of space while the other three interactions – electromagnetic, weak and strong – can determine modifications in the "vibrations" of the elementary grains of the region of space into examination. It is in this sense that the special state of space determines the appearance in certain quanta of space of a discrete quantity of entropic energy and thus of a material particle.

As regards the interpretation of matter, according to the view suggested by the authors of this book, one can say that the fundamental sources of the quantum information emerging from the processes of the three-dimensional quantum vacuum are the vibrations at appropriate frequencies (characteristic of material particles). In synthesis, the following perspectives are opened by this model. On one hand, we can say that this model can give results similar to standard quantum theory, in the sense that the quantum states of the particles into consideration coincide with the correspondent states predicted by standard quantum physics. On the other hand, at the same time, this model introduces these new simple elements: the idea that space is timeless and has a granular structure at the Planck scale and the fact that the state of each subatomic particle can be always seen as the consequence of the vibrations of one or more quanta of space at appropriate frequencies (emerging from a timeless three-dimensional quantum vacuum whose most fundamental properties are the energy density and RS processes of creation and annihilation of quanta).

Inside the theory of everything here suggested, the state of a subatomic particle (as we know it from quantum theory) can be seen as a consequence of the vibrations of one or more quanta of space with appropriate frequencies (characteristic of the particle into examination). Also the wave behaviour of the subatomic particles, the quantum waves associated with the material particles, can be seen as effects of the vibrations of one or more quanta of space at appropriate frequencies. The vibrations of the quanta of space at appropriate frequencies create the appearance of material particles (in the sense that, because of them, these quanta of space become seat of a discrete quantity of entropic energy) and, at the same time, determine their wave behaviour, the quantum waves associated with them. And, on the ground of the equations of this model, one can say that these waves guide the corresponding particles, through the action of quantum potential, in the different points of the timeless space: the entropic energy is transferred by the quantum potential among the different quanta of space composing the trajectory described by the subatomic particle. Let us consider, for example, the behaviour of the electron in the hydrogen atom. The role of the quantum potential is to transfer, to guide an opportune quantity of entropic energy among the different quanta of space composing the trajectory described by the electron in an energetic level of the hydrogen atom (and which, of course, emerge from opportune changes of the three-dimensional quantum vacuum energy density and from opportune processes of creation or annihilation). In other words, we can say that there is a sort of correspondence between quantum potential and the appearance of entropic energy in the different points of the timeless space. The role of quantum potential is to transfer entropic energy among the quanta of space occupied by a subatomic particle during its motion, to guide this entropic energy among the different quanta of space composing the trajectory described by the particle in the timeless space.

The theory of everything in the timeless space allows us to provide an interesting interpretation of the fundamental starting-point of Bohm's theory, namely the fact that every subatomic particle can be seen as a structure constituted by a wave and a corpuscle with the wave that acts as a pilot wave. In fact, the perspective is opened that the waves associated with subatomic particles and that guide them during their motion can be seen as an effect of the vibrations, with appropriate frequencies, of the quanta of space that originate them.

Finally, it is important to underline that, in the timeless theory of everything, the spin of a subatomic particle is associated with (and derives from) a quantum number of the rotation-orientation of the quanta of space characterizing the region under consideration. Of course, some of the ideas here illustrated are in studying process and further research will give you more information about these topics.

2.6. Quantum Potential as "Special State of the Timeless Space in the Presence of Microscopic Processes"

In Bohm's pilot wave theory, the movement of a subatomic particle is determined by the sum of a classical potential and a quantum potential. On the other hand, in the model of a timeless granular space suggested by the authors, the motion of a subatomic particle is tied to the interaction of an opportune quantity of entropic energy with the various quanta of space composing the trajectory described by the particle into consideration (and this interaction is produced by the quantum information contained in these quanta of space and, in particular, in their vibrations at certain frequencies, which emerge from the RS processes of creation and annihilation and changes of the energy density of the fundamental timeless three-dimensional quantum vacuum). According to this model, the vibration at appropriate frequencies of one or more quanta of space determines the interaction of a certain entropic energy with these quanta of space and the appearance in that region of a subatomic particle; this vibration produces also a quantum wave that guides the particle during its motion, making it appear in different points of space.

The connection between the timeless interpretation of the law of motion in Bohm's pilot-wave theory and the interpretation of a subatomic particle in quantized space is clear and physically compelling. As we have shown in the previous paragraph, in our theory of everything, it is the vibration of certain quanta of space at appropriate frequencies that determines the appearance of a particle in space and creates the wave which guides the particle during its motion. Therefore, the motion, the stream of changes to which a particle is subjected in space is itself tied to the vibration at appropriate frequencies of the quanta of space composing its trajectory. This implies that quantum potential itself is determined, in short, by the vibration at appropriate frequencies of some quanta of space.

According to the theory of everything in the timeless space, the following idea can be therefore proposed. There is someway a connection, a correspondence, between the entropic energy movement among the various quanta of space, these elementary grains of space – movement that, as said before, materializes a particle making it appear in the different points of space – and the action of the total force that produces the motion of that particle, as we know it from Bohm's pilot wave theory. The movement of a particle under the action of the total force (classical + quantum) can be seen as the effect of the entropic energy transfer among the different quanta of space in which the trajectory described by the particle can be decomposed. In other words, one can think that the total force (classical + quantum) which acts on a particle represents the immediate effect, the immediate manifestation of the entropic energy shifting between the various quanta of space composing the trajectory described by this particle [11, 12].

In this way, a new interpretation of the quantum potential can be proposed: the idea that the quantum potential has origin from the fundamental level of reality, in which physical space is timeless, is composed by a net of quanta of space having the size of Planck's length (which represent its elementary grains) and in which one assumes that subatomic particles appear in virtue of the interaction of entropic energy with one or more quanta of space (which emerges from the RS processes of creation or annihilation of quanta from the three-dimensional quantum vacuum). This new interpretation opens the possibility that there is a possible link between quantum potential and granular structure of physical space at the Planck scale as a consequence of the fundamental processes of the three-dimensional quantum vacuum namely the RS processes. It implies in fact a correspondence between bohmian quantum potential and the entropic energy shifting among the quanta of space (responsible of the materialization in different points of space of an elementary particle): as regards microscopic processes, space assumes the special "state", represented by the quantum potential, in consequence of the entropic energy shifting between various quanta of space, and therefore in consequence of the vibration of such quanta of space at appropriate frequencies (produced just by opportune RS processes of the three-dimensional quantum vacuum). In short, the movement of atomic and subatomic particles is determined by the entropic energy shifting among the various quanta of space (caused by the vibration of these quanta of space at certain frequencies emerging in turn from opportune RS processes from the three-dimensional quantum vacuum) and it is as a consequence of this energetic shifting that space assumes the special "state" represented by the quantum potential [11, 12, 35- 38, 40-42].

The interpretation of quantum potential as the special "state" of the timeless space in the presence of microscopic processes, can be also seen as a natural consequence which derives from quantum non-locality. Non-locality provides to us another element for which this new interpretation of quantum potential appears permissible and coherent from the physical point of view. As we have illustrated in chapter 1.8, the instantaneous communication between two subatomic particles separated by big distances, in Bohm's version of quantum mechanics, is explained by the quantum potential. But at the same time, as we said in chapter 1.7, non-locality of quantum phenomena can be also seen as an effect of the timeless space. That is to say: one can think that it is the timeless space to transmit the information between two particles, before joined and then removed and carried at big distances one from the other, to let these two particles communicate instantaneously. Hence we are presented immediately with the possibility that there is a sort of correspondence between quantum potential and the timeless space, in particular that quantum potential can be just interpreted as the special "state" of the timeless space in the presence of microscopic processes, and thus of quantum particles. When one takes into consideration an atomic or subatomic process (such as for example the case of an EPR-type experiment, of two subatomic particles, before joined and then separated and carried away at big distances one from the other), space assumes the special "state" represented by quantum potential, and this allows an instantaneous communication between the two particles under study.

It is important to point out, finally, that this new interpretation of quantum potential as "special state of timeless space in the presence of microscopic processes" is substantially compatible with the idea, originally proposed by Bohm and Hiley in 1984, to consider the quantum potential as an "information potential" and allows to bring new light on it. In fact, saying that the quantum potential contains an active information about the region of the experiment and that this information manifests itself in the movement of the particles is equivalent to saying that it includes the whole of features that the region of experiment presents, and that such features determine just the movement of the particles. Quantum potential contains an active information just because it represents, at the deepest level, the state of the timeless space in the presence of microscopic processes: its way of acting emerges from what happens at a deeper level of reality, namely from the three-dimensional quantum vacuum, it is just from the changes of the quantum vacuum energy density and from the RS processes that its feature to contain an active information as regards the region of the experiment arises.

After having illustrated the new interpretation of quantum potential inside the timeless theory of everything, in order to understand how it allows us to throw new light on the behaviour of subatomic particles, let us see now in detail some fundamental experiments of atomic physics.

2.7. THE DOUBLE-SLIT INTERFERENCE

The first important experiment that we face is the double-slit interference. It is the classic experiment in which we have a beam of electrons which are sent, one at a time, to a double-slit, and then are revealed on a screen, where an interference pattern is observed. As it was underlined by Feynman, the understanding of what happens in the double-slit experiment takes part a little of the heart of the problems of quantum mechanics. In fact, standard quantum mechanics (Copenhagen interpretation) is hardly able to let us understand what happens when electrons are sent toward a double-slit apparatus, to let understand why these electrons reproduce the interference pattern. In the standard version of quantum mechanics, we can only say what is the probability that the electron finishes in a certain point of the screen, but we cannot follow its motion, we cannot speak of the electron's trajectory after it has passed through the slits, in short we cannot provide a lecture in visualizable terms of this process (see, for example, the book [43]).

Instead, Bohm's theory (on the ground of the calculations made by the British physicists Philippidis, Dewdney and Hiley in 1979 [44]) allows us to explain what happens in the double-slit experiment in agreement with a principle of causality. One can keep the concept of well-defined trajectories of the particles, and the final position of the particle on the screen allows us to deduce the slit through which it has passed. Bohm's theory shows how an accurate calculation of quantum potential for the double-slit experiment can generate the interference pattern without leaving the concept of well-defined trajectories of the particles. By analysing double-slit experiment on the ground of this theory, namely, one obtains the following results. Each single particle follows a precise and calculable trajectory which comes from one slit or the other and this aggregate of trajectories produces the requested interference figure and, at the same time, shows that the final position of the particle on the screen permits to deduce through which slit it has passed; besides, if the distribution of the arrivals on the two slits is uniform, the action of quantum potential assures that the probability density of presence of the particles remain equal to the modulus square of the wave-function in all the following instants [45, 46].

Now, utilizing the results displayed in the previous paragraphs about the link between the quantum potential and a timeless background, one can bring new light on double-slit experiment and provide a significant justification – at least in an interpretative way – of how the particles emitted by the source move in the region between the slits and the screen, of their resulting trajectories (which are considered "strange" by some physicists of Copenhagen interpretation).

The trajectory described by each subatomic particle can be interpreted as the result of the energy transfer between some quanta of space: the quanta of space occupied by a particle during its motion – as a consequence of opportune changes of the quantum vacuum energy density and of opportune RS processes from the three-dimensional quantum vacuum – vibrate at appropriate frequencies and assume therefore a discrete quantity of entropic energy, perceivable to our senses. In consequence of the energy shifting between the different quanta of space composing the trajectory described by each particle coming from the beam, the region of space of the double-slit experiment assumes the special state represented by quantum potential. If in Bohm's original theory the trajectory described by each particle in the double-slit experiment is tied to quantum potential, now introducing our interpretation of quantum potential as "special state of timeless physical space in presence of microscopic processes", one can say something more (and provide perhaps a significant adding explanation of the origin of the trajectories of the particles). In fact quantum potential, in our interpretation, draws its origin at the deepest level of reality, in which physical space is timeless, endowed with a granular structure, and the materialization of each subatomic particle in the different points is tied to an entropic energy transfer – determined by opportune RS processes of creation and annihilation of the timeless three-dimensional quantum vacuum – among the quanta of space composing its trajectory. The region of double-slit assumes the special "state" represented by quantum potential (whose action determines, in Bohm's theory, one determined trajectory for each particle of the beam), in virtue of what happens at a deeper level of reality, i.e., in virtue of the interaction of entropic energy with certain quanta of space (caused by the vibration at certain frequencies which in turn is produced by more fundamental opportune RS processes of creation and annihilation of the timeless three-dimensional quantum vacuum). It is the energy shifting between certain quanta of space that produces the movement of each particle of the beam that is responsible of its following appearance in different points and thus explains the particular way in which quantum potential acts. It is just in this deeper level of reality that the action of the quantum potential emerges.

Therefore, if one utilizes this interpretation of quantum potential as "special state of timeless physical space in presence of microscopic processes", the aggregate of trajectories described by the electrons in double-slit experiment must not be considered devoid of sense, because they are produced by the events concerning the most fundamental level of reality: the following appearance of a particle in different points of space is tied to the interaction of entropic energy with certain quanta of space of the region of space into examination, which is determined by the vibration of these quanta of space at appropriate frequencies (emerging from opportune RS processes of the three-dimensional quantum vacuum). Practically, on the basis of this new interpretation of quantum potential, the following perspectives are opened. When a single electron arrives on the slits, it is the particular information contained in the quanta of space composing that region of space, to determine what will happen to such electron successively (i.e., after the passing through the slits). The quantum information contained in the quanta of space, when the first electron is sent to the slits, is characterized by some properties which derive from fundamental RS processes of the three-dimensional quantum vacuum, and it is just this quantum information which determines an appropriate action of the quantum potential on that electron, and so makes energy appear in certain points instead of others, which produces thus a determined trajectory for that electron. Then, when the second electron arrives on the double-slit, it is permissible to think that the quantum information contained in the quanta of space composing that region of space is different (because of the characteristics of the fundamental RS processes from which it emerges). This different quantum information contained in the quanta of space determines a "different" action of quantum potential, allowing the electron, this time, to follow another trajectory (different from that of the first electron): this time, there is the vibration, and thus the creation of entropic energy, in different quanta of space (as regards those characterizing the state of space at the beginning of the experiment). These considerations can be then extended to all the particles which are sent, one at a time, toward the double-slit. This means that, on the ground of the interpretation of quantum potential as "special state of timeless physical space in presence of microscopic processes", it is possible to provide a significant justification – at least at an interpretative level – of the trajectory described by each particle of the beam.

2.8. TUNNELLING

The second experiment we take in consideration is the overcoming of a potential barrier by tunnelling. If one sends a beam of particles, one at a time, towards a potential barrier, one observes experimentally that a given particle has a finite probability to overcome the potential barrier also when its initial energy is less than the barrier. In the standard version of quantum mechanics, it is not possible to provide a reading in visualizable terms of this typical quantum phenomenon. Instead, on the ground of the results of Dewdney, Bohm's theory can provide a clear explanation, in terms of motions of single particles, of what happens when a single particle reaches the potential barrier. This theory shows that some particles are reflected and some manage to overcome the barrier, according to their initial position. In particular, applying Bohm's theory to this problem, one finds that in the barrier region the quantum potential is negative, and this determines an effective lowering of the barrier, which therefore can be overcome by a particle (if it has got sufficient energy) without violation of energy conservation law [47, 48].

The interpretation of quantum potential as "special state of timeless physical space in presence of microscopic processes" proposed by the authors of this book allows us to bring new light also to this experiment about tunnelling. In virtue of the link between quantum potential and the energy shifting between the quanta of space composing a region of space, it is permissible to think that the negative value of quantum potential in the barrier region is tied to the particular "quantum information" contained in the quanta of space composing such region (which emerges from opportune RS processes of creation/annihilation from the timeless three-dimensional quantum vacuum). Besides, the fact that the particles having bigger initial energy (and so speed) have bigger probability to overcome the barrier, can be explained taking into account that these particles pass more rapidly from one point to another of their trajectory as a consequence of the specific quantum information contained in opportune quanta of space. This means then that for these particles endowed with bigger speed, in virtue of the peculiar quantum information of the quanta of space, the energy transfer between the different quanta of space composing their trajectory happens more rapidly than that for the particles having smaller initial speed: just this determines, for particles endowed with bigger initial speed, a bigger probability to overcome the barrier.

In conclusion, on the ground of the idea that, at the most profound and fundamental level of reality, the motion of each subatomic particle in space is tied to the entropic energy shifting between the various quanta of space composing its trajectory (as a consequence of the quantum information of these quanta of space emerging from opportune processes of creation/annihilation from the timeless three-dimensional quantum vacuum), it is possible to provide a clear reading, a convincing explanation – at least at an interpretative level – of what happens in tunnelling experiment, it is possible to understand clearly why a particle endowed with sufficient energy can overcome the potential barrier.

2.9. THE FUNDAMENTAL INTERACTIONS IN THE TIMELESS SPACE

On the basis of the postulate 3 of the theory of everything, universe is a self-renewing system where matter and physical space are in a permanent dynamic equilibrium. According to the standard cosmological model, the fundamental interactions existing in nature (gravitational, electromagnetic, weak and strong) appeared in different stages of the evolution of the universe. In the standard cosmological model, the evolution of the universe can be seen as a succession of formations of aggregated structures at increasing stages: before all, the quarks give place to the common nuclear forces and thus to nuclei; successively, we have the formation of atoms because of the electromagnetic forces acting between nuclei and electrons; then, in a following stage, we have the formation of molecules in virtue of the residual Van der Waals forces acting among the already formed atoms; finally, because of the specific characteristic of carbon, we have the formation of the multi-molecular structures, of tens, thousands, millions of atoms, which lead to the cellules, to the pluricellular beings, and finally to the whole biology outline.

Universe here is a timeless phenomenon in the sense that in the universe there is no “before” and no “after”, it is only “now”. The only universe that exists is the one we can experience. With clocks we measure the speed of material change which depends on the speed of an inertial system (as to another inertial system) and on the quantum vacuum energy density where material change occurs and is measured. Duration of irreversible material change is its “rational description” based on the concept of linear time and does not exist in the physical reality itself, at a fundamental level.

In this view of the universe, one can put in discussion the idea that the fundamental interactions appeared at different stages, that before nuclear forces appeared, then electromagnetic forces appeared and so on. In the timeless universe all the different force fields appeared at different periods only in the sense of a numerical order. In a timeless universe whose only physical reality is the “now”, all the interactions can be always present. They represent the different “states” of space, the “characteristics” of the various regions of the universe.

According to the postulate 4 of the timeless theory of everything, all the fundamental interactions and physical fields represent the “states” of space, the special environment situations in the presence of certain material particles. The fundamental interactions and physical fields are special states of the timeless space determined by the existence of material vibratory states in various quanta of space (and thus by the peculiar quantum information contained in opportune quanta of space which emerges from opportune RS processes of creation/ annihilation from the three-dimensional quantum vacuum). If a quantum of space, vibrating at a certain appropriate frequency, assumes a discrete quantity of entropic energy, thus becoming material quantum, a modification in the properties of the surrounding region of space (for example in the frequencies of vibration of the surrounding quanta of space) can be generated. The fundamental interactions, the different physical fields can be interpreted just as the modifications in the properties of a region of space caused by the presence in that region of appropriate particles (and thus determined by the quantum information of opportune quanta of space). These modifications depend on the features of the particles and in turn influence the motions of the closest particles, determining their most favourable trajectories. In other words, one can say that the presence of entropic energy in one or more quanta of space determines a variation of the state of the surrounding quanta of space, which can so have more or less disposition to the energy transfers.

If all physical fields are always present because they represent special states of the timeless space, in the model here analysed the state of space, as describes a granular entity, can be represented through a sort of generic "linear superposition" of the different physical fields, namely gravitation, electromagnetism, strong nuclear force, weak nuclear force.

As known, each force field is characterized by its own relative strength parameter, operates over a specific distance or time scale, has its own range, its own typical lifetimes for decay and, finally, is mediated by its particular fundamental bosons. One can argue that it is just the particular values of these properties and quantities to cause, to determine the "collapse" of the generic state of space (as a sort of quantum superposition of the different physical fields) in the force field characterized by those values of properties and quantities. In particular, it is the particular value of the range of the modifications induced in the properties of space and the particular features (i.e., vibrations) of the bosons transmitting the interaction which determine the "collapse" of the generic state of space in the force field characterized by those properties. Our perception of distinct physical fields is therefore tied to the particular frequencies of vibration of quanta of space (which determine the appearance of certain bosons) joined to the particular value of other properties or parameters, such as the range of the interaction (that we interpret as the range of the modifications determined by that interaction in the properties of space) [35, 36, 38].

A particular distinct consideration concerns the quantum potential which can be seen as the state of the timeless space when it is characterized by microscopic processes, and therefore when we have problems in which Planck's constant assumes an important role (and it is just the presence of this term to produce non-local correlations between quantum particles, which instead were absent in the macroscopic world of classical physics). The term represented by quantum potential can therefore appear simultaneously with each of the other terms, just to indicate the fact that we are in the quantum regime (and we have non-local correlations among the particles). One can say that it is thus the quantum potential the ultimate entity that can account of the non-locality of the fundamental physical processes and, at the same time, that can be considered the simplest way to introduce the non-locality in each mathematical approach describing the quantum foams. Besides, as we have seen before, it is important to underline again that, in this model, the role of quantum potential is to guide a discrete quantity of entropic energy among the different quanta of space composing the trajectory described by a subatomic particle during its motion.

2.10. About the Wave Picture of the Timeless Theory of Everything

In 1950 Einstein showed that space must have a property to connect particles among them, allowing thus the transmission of a force, of an interaction from one particle to another. According to Einstein, the combination of the idea of a continuous field with the idea of discontinuous points in space appears inconsistent: the concept of discrete particle cannot be present in a coherent field theory. So, according to Einstein, also without including gravitation, Maxwell's electrodynamics cannot be considered a complete theory. In agreement with the point of view of Greek philosophers and mathematicians of the time who asserted that all matter and motion derive from one substance, Einstein rejected the conception of matter as constituted by discrete particles (because forces and properties of nature do not agree with the discrete particle) and tried to represent matter as spherical fields in space-time.

An interesting theory that can face these questions claimed by Einstein indeed already exists: it is the theory of "Wave Structure of Matter" developed by the American physicists Wolff, Mead and Haselhurst [49-52]. In this theory, a wave structure replace completely the particles as discrete material points, in agreement with what Einstein wanted: to find that a subatomic particle, such as the electron, does not exist as discrete material point, to remove Bohr's interpretation of the wave function and to assure that God does not play dice, eliminating therefore all the paradoxes of quantum mechanics. The theory of Wave Structure of Matter is practically based on the following two ideas: 1. that quantum waves exist in space and are solutions of a wave scalar equation, and 2. that the waves coming from all the particles of the universe combine their intensities to form the wave medium represented by space (and, therefore, that it is the waves that create space).

In the theory of Wave Structure of Matter the waves of each particle are intermingled with the waves of other particles and all contribute to the density of the space medium: this implies that in this theory the density of space depends on the distant matter.

Now, we can observe that the theory of everything in the timeless space analysed in this book allows us to make some important progresses with respect to the picture proposed by the theory of the Wave Structure of Matter. While in the theory of the Wave Structure of Matter, there is not indeed a fundamental medium, in the sense that the background space depends on the waves coming from all the particles of the universe, instead in the picture suggested by the timeless theory of everything there is indeed a fundamental medium from which everything derives: it is the timeless space characterized by the basic frequency of vibration, in other words it is the timeless three-dimensional quantum vacuum characterized by the maximum value of the energy density. Therefore, in our theory of everything, contrary to what occurs in the theory of the Wave Structure of Matter, the existence of the background space does not depend on matter because the background constituted by elementary grains vibrating at the basic frequency (which corresponds to the maximum value of the quantum vacuum energy density) is timeless. It is matter that depends on the states characterizing the timeless space. In the view here suggested, it is just the vibrations of quanta of space at the frequencies characteristic of subatomic particles – emerging from opportune RS processes of creation/annihilation from the three-dimensional quantum vacuum – which give place to the quantum waves associated to the material particles. The wave behaviour of the subatomic particles arises from the vibrations of quanta of space constituting them.

One can also say that the waves associated with the material particles are created by the states characterizing the timeless space. The quantum waves of matter arise from the particular environment situations characterizing given regions of space. The peculiar environment situation characterizing a given region determines the lowering of the frequency of one or more quanta of space from the basic frequency (that characterizes empty timeless space) to the frequencies characteristic of material particles; and, then, it is just these vibrations at the frequencies characteristic of the particles that determine the quantum waves associated to those material particles.

These quantum waves can be interpreted both in the standard sense (as mathematical tools to compute certain probabilities) and in more "realistic" senses (like in Bohm's pilot-wave theory or in the theory of Wave Structure of Matter, for example). We emphasize however that we prefer to interpret them in a realistic sense. We suggest, in particular, that these quantum waves can be interpreted like in Bohm's pilot wave theory. In our theory of everything, as regards the elementary grains of space vibrating at the frequencies of material particles, the appearance of a mass is related to the quantum potential: the role of the quantum potential is to transfer a discrete quantity of entropic energy (associated with a subatomic particle) among the different quanta of space composing the trajectory described by the subatomic particle in space. The quantum waves associated with the material particles guide the corresponding particles during their motion, through the action of quantum potential, in the regions where the quantum wave function is more intense.

These quantum waves, determined by the vibrations at appropriate frequencies of quanta of space, confer a wave structure to the timeless universe. Therefore, universe turns out to have, at the same time, a granular structure, given by a network of quanta of space (just like in loop quantum gravity), but also a wave structure (in analogous way to the picture provided the theory of the Wave Structure of Matter), in virtue of the vibrations of the quanta of space occupied by matter, namely the quantum waves associated with the subatomic particles. And, as we have already said, these quantum waves, like in Bohm's pilot wave theory, guide the corresponding particles during their movement (in the regions where the wave function is more intense). As a consequence of this "realistic" interpretation, the quantum waves contribute to make all particles interact, link all particles. In this scheme, we have therefore an image of physical world which, from the philosophical point of view, is compatible with some aspects of the Wave Structure of Matter. In fact, both in the theory of the Wave Structure of Matter and in the theory of everything proposed here, we have quantum waves that arise from peculiar regions of the universe and that link the different particles between them. The most important difference between the two approaches, from the philosophical point of view, lies in the fact that while in the theory of the Wave Structure of Matter the space medium is not fundamental but derives from the waves of all the particles of the universe, in the theory of everything suggested here there is really a fundamental medium, constituted by the timeless background characterized by the basic frequency of vibration (namely by the maximum value of the quantum vacuum energy density). Moreover, the picture of the universe provided by the timeless theory of everything, above all, appears consistent under Einstein's point of view (because here the waves create just a connection among all the particles, and as we have told, all this can be considered an effect of the vibrations of opportune quanta of space subjected to appropriate "states" of the background space – and which emerge from opportune fundamental RS processes of creation and annihilation characterizing the timeless three-dimensional quantum vacuum).

One can say, like in the theory of Wave Structure of Matter, that these waves form and build up the space medium – which connect the particles among them. But here, contrary to the theory of Wave Structure of Matter, it is not the waves associated to all the particles of the universe that combine themselves to give place to the space medium because the space medium characterized by the basic frequency is timeless and its existence does not depend on matter, on the motion of particles; instead, it is lowering of the frequency of vibration of peculiar quanta of space (corresponding to peculiar RS processes of creation/annihilation from the quantum vacuum) that creates material particles and thus the quantum waves associated with the subatomic particles. It is the motion of a peculiar vibratory state among opportune elementary grains of space (corresponding to the occurrence of peculiar RS processes of creation/annihilation from the quantum vacuum) that create the appearance of material particles and therefore of the quantum waves associated with them.

In the theory of Wave Structure of Matter, space is a quantum wave medium created by the waves coming from each particle of the universe: the matter of the universe creates the wave medium in all space and the waves coming from each particle create the space medium. Instead, in the view suggested by the authors of this book, this symmetric, reciprocal effect is not possible, because matter derives from physical space (in the sense that it is produced by the special state of a given region of space, which has the effect to change the frequency of vibration of quanta of space from the basic frequency to a frequency characteristic of a material particle, and this vibration determines in turn the quantum waves associated to material particles) but the background characterized by the basic frequency – namely by the maximum value of the quantum vacuum energy density – cannot be derived from matter because is fundamental.

So one can also say that our model, as far as its conceptual and philosophical foundations are concerned, cannot be completely considered compatible with the view of space of the theory of Wave Structure of Matter. But it can be compatible with some important ideas of Bohm's pilot wave theory. It can provide an interesting interpretation of the main starting-hypothesis of Bohm's pilot wave theory, namely the idea that every subatomic particle must be thought as constituted by a wave and a corpuscle at the same time, with the wave that guides the corpuscle during its motion. In fact, in virtue of what we have said in chapter

2.5 and 2.6, one can propose that the waves associated with subatomic particles and that guide them during their motion can be considered the effect of the vibrations – at appropriate frequencies – of the quanta of space that are occupied by these peculiar vibratory states (and therefore by the peculiar corresponding RS processes of creation/annihilation from the three-dimensional quantum vacuum).

For example, the appearance of the electron in a given region of the timeless space is determined by the interaction of energy in the "entropy state" with one quantum of space in the elementary grains of space that compose the trajectory followed by the electron. This interaction of entropic energy with one quantum of space in the points of the trajectory of the electron is produced by the motion of an appropriate vibratory state among the quanta of space in which the electron appears (and thus by the occurrence of particular RS processes of creation/annihilation from the three-dimensional quantum vacuum). It is the motion of an appropriate vibration of an elementary grain of space among different quanta of space (and thus the occurrence of particular RS processes of creation/annihilation) that produces the wave associated with the electron, and this wave guides the electron, through the action of quantum potential, in the regions where the wave-function is more intense, making it appear in different points of space: the role of quantum potential is to transfer a discrete quantity of entropic energy (associated to material quanta) among the different quanta of space composing the trajectory described by the particle.

2.11. A New Interpretation of Superstring Theory and of Loop Quantum Gravity in the Timeless Space

The ideas developed in this chapter and the previous one as regards a timeless arena of the universe in which time is only a measuring system of the numerical order of material change, allow us to provide a new key of reading and interpretation of superstring theory. Does a string vibrate in a physical reality of space-time, or a string vibrates only in a timeless space where time is not a physical reality that flows on its own? This question can also be turned as: is time part of space or is time only numerical order of string vibration that we measure with a clock? In virtue of the results mentioned in chapter 1, the most appropriate answers to these fundamental questions are clear. The concept of a temporal dimension that flows on its own in the universe must be replaced with the concept of numerical order of material changes occurring in a timeless space. Hence, a new understanding of strings is opened: at the fundamental level of reality of

superstring theory, time represents only the numerical order of a string vibration. The frequency γ of a given string is determined by the numerical order of its vibration: $\gamma = \frac{1}{t_n}$, where t_n is numerical order of vibration. Therefore, a new suggestive interpretation of superstring theory can be proposed in which all what we perceive in the physical world (matter, force fields with their properties) turns out to be the manifestation of specific notes that string play, of specific patterns of vibration of fundamental strings just because these fundamental entities can be characterized by different numerical orders of vibration. It is the numerical order of vibration of the string the fundamental reality that determines the characteristics of all what we see in the physical world. Superstring theory can provide a unitary theory of matter and interactions on the basis of the fact that the strings are endowed with a spectrum of possible vibrations which are characterized by several possible numerical orders. The numerical order of vibration of a string can assume several values and this fact produces the appearance in the physical world of the different particles with their own properties.

In analogous way, the timeless theory of everything suggested in this chapter allows us to provide a new interesting reading and interpretation also of loop quantum gravity. The history of spin networks obtained with a reading on a physical clock does not happen in an absolute sense, but occurs only in the sense of the numerical order of events regarding these spin network states. In loop quantum gravity an absolute time that flows on its own does not exist just because clocks are measuring systems of the numerical order of changes in the gravitational field. We can say that the timeless theory of everything suggested in this chapter allows us to go beyond and improve some relevant ideas already considered by loop quantum gravity. In fact, also in the model here suggested, just like in loop quantum gravity, time has not a fundamental reality, the duration of events has not an objective existence on its own, and physical space has a granular structure, is composed by elementary grains of the size of Planck scale. Moreover, the timeless theory of everything suggested here can develop in an appealing way the idea of the granularity of space by introducing the interesting perspective to characterize each elementary grain of space with a specific frequency of vibration (which derives from opportune RS processes of creation/annihilation from the three-dimensional quantum vacuum). Thus, under this point of view, the theory proposed by the authors of this book takes a fundamental idea of superstring theory and puts it in a picture of space similar to that of loop quantum gravity.

If we analyze the fundamental ideas and the pictures of the world of superstring theory and loop quantum gravity (and somewhat also of the theory of the Wave Structure of Matter, illustrated in chapter 2.10), it seems evident that these theories determine pictures of the universe that, at a first sight, do not seem compatible one with each of the others. Superstring theory, loop quantum gravity (and somewhat also the theory of the Wave Structure of Matter) provide different answers as regards the nature of space.

Instead, on the basis of the treatment made in this book, the timeless theory of everything suggested in this chapter allows us to unify in a certain way the pictures of the universe proposed by loop quantum gravity and superstring theory (and also of the theory of the Wave Structure of Matter, taking into account what we have said in chapter 2.10). The view suggested by the timeless theory of everything opens the possibility to provide a unitary and holistic picture of the universe and to throw a new light on vibratory, granular and wave nature. This implies that, from the philosophical point of view, the picture of the universe determined by the timeless theory of everything can be considered more convincing than those of string theory, of loop quantum gravity (and of the theory of the Wave Structure of Matter).

In virtue of the results mentioned in chapter 1, the timeless theory of everything suggests the starting-point perspective that space-time is not the fundamental physical arena of the universe. On the other hand, from the perceptive, experimental point of view, there is no evidence whatsoever to support the view that space-time exists as the basis of a fundamental physical reality. Space-time cannot be considered a primary physical structure.

The timeless theory of everything develops and realizes in a suggestive way what Einstein wanted as regards a unitary view of the universe and what Einstein already underlined as regards the fact that it makes no sense to think about space-time as an independent physical entity. The timeless theory of everything suggests that our experience of space is based on the observation of material objects while our experience of time derives from the changes and movements of matter into the universe. Time is a mathematical coordinate which describes the motion of material objects. Time exists only as a numerical order of material changes and motions occurring in a timeless gravitational space and enters into existence when we measure it through clocks. With clocks we measure the numerical order of material changes that run into a timeless universe. According to the timeless theory of everything the fundamental arena of the universe is a timeless physical quantum-gravity space – having a granular structure at the Planck scale – which emerges from the timeless three-dimensional quantum vacuum and the wave function characterizing each elementary grain of space is its fundamental element.

Both granularity and wave nature turn out to be intrinsic features of space inside this theory. Practically, all what we observe and perceive can be seen as the effect of specific frequencies of vibration of the elementary grains of space of Planck scale. We can say that we are immerged in an infinite vibratory field, in a real cosmic symphony: a peculiar vibration of the elementary grains corresponds to each entity and these frequencies are the elements that determine the evolution and thus, under this point of view, the universe is a conscious phenomenon. It is for the human being to realize that his/her nature is in harmony with the order of the universe.

2.12. Towards a Holistic View of the Physical World

The view suggested by the timeless theory of everything illustrated in this chapter can provide a unitary and holistic description to the physical universe. On the basis of this theory, our perception of the different physical entities is linked to the vibrations which characterize the elementary grains of the timeless space (and which emerge from opportune RS processes of creation/annihilation from the three-dimensional quantum vacuum). Just like in superstring theory, all we perceive in the universe is a consequence of specific frequencies of vibration and, just like in loop quantum gravity, the elementary constituents are quanta of space of the size of Planck scale and, just like in the theory of the Wave Structure of Matter, all the elementary grains of space are characterized by a wave nature. The theory of everything suggested in this chapter introduces therefore an appealing unification of the fundamental ideas contained in different relevant theories of physics as superstring theory, loop quantum gravity and the theory of the Wave Structure of Matter.

The most important perspective introduced by this theory lies just in the mapping of the universe as a whole of grains endowed with appropriate frequencies of vibration. The different material and non material properties of the universe are determined by its granular and wave nature and therefore by the vibrations of its most elementary constituents. According to view proposed here, the frequency of vibration of the quanta of space can be thus considered as the fundamental element which determines the characteristics of the entities present in the universe. We can say that we live practically in a sea of frequencies: our perception of the different entities is the consequence of the different frequencies of vibration of the most elementary constituents of the universe.

It is easy to imagine the universe as divided into its parts. But it is not easy to imagine the mutual motion of these parts to create space. The view suggested here allows us to realize that the different physical entities do not move in space but, being parts of space, they perform a mutual motion that is the consequence of the peculiar vibrations of their most elementary constituents (which in turn derive from the occurrence of specific processes of creation/annihilation of material quanta from the timeless three-dimensional quantum vacuum). Every selected part of space, by being composed by certain quanta of space characterized by their specific properties such as frequencies of vibration contains information and thus determines the behaviour of a certain part of the world we perceive. According to the standard view, every particle, object or electromagnetic field moves in a background space. Instead, the view proposed here introduces the idea that the timeless physical space, dragged by a particle or any material object, moves towards the rest of the timeless space in virtue of the specific vibrations of their most elementary constituents. Moving objects are mutually moving parts of the timeless space. Moreover, the mutual motion of the different physical entities is possible only thanks to the mutual connections between the elementary grains of space that compose them. These connections are also the parts of space and are essential. If the part is separated from the whole, it does not leave this whole, but remains jointed with it by its universal connection. The part obtains its relative independence only thanks to its universal connection, which allows the part to obtain the possibility for its relative motion and to remain connected with the whole at the same time. The connection between the elementary grains of space is also a characteristic of the timeless space. So there are no differences between part and connection. Every connection is a part of space and every part of space is a network of elementary connections. In the timeless view suggested by the authors, everything is connected with everything, as regards the information contained in the elementary grains of space. Every separated part is connected with all the parts of the universe as regards their fundamental properties. In virtue of the vibrations of the most elementary grains of space, a universal connection of everything with everything creates the general unity of the universe. The universe can be considered as a network of non-local connections and "perceives" all of its parts and, on the other hand, every elementary quantum of space "perceives" the whole universe. Quanta of space with their specific properties (in particular, the frequencies of vibration) – which emerge from the RS processes of creation/annihilation characterizing the timeless three-dimensional quantum vacuum – not only exist but represent the substance of the universe. It means that the universe is always universally interconnected in virtue of the peculiar properties of its most elementary constituents.

Moreover, it is interesting to observe that our timeless theory of everything allows us to make some relevant considerations about the development of scientific knowledge. From a general point of view, a physical theory can be defined as a conceptual structure that we develop and use to organize, read and understand the world, and make predictions about it. A successful physical theory is a theory that manages to reach these aims in a consistent and effective way. However, each new theoretical scheme has to take account of what previous theories have already reached. As Rovelli rightly underlined scientific thinking is not a static entity but is in constant evolution and reorganization: science is itself the evolution process of thinking [18]. There is no reason to believe that our understanding of the world has not to be in constant evolution. At each step the best that we can say about the reality of the world is what we are saying in that moment. The fact that successively we will better understand the world does not make our understanding less valuable or less credible. Searching for a fixed point on which to rest our restlessness is naïve, useless and counterproductive for the development of knowledge. It is only by believing our insights and, at the same time, questioning our mental habits, that we can go ahead. This process of cautious faith and self confident doubt is the core of the scientific thinking. Science is the human adventure that consists in exploring possible ways of thinking of the world. Being ready to subvert, if required, anything we have been thinking so far can be considered one of the best of human adventures. The considerations made in this book about a timeless background as ultimate arena physics (and thus about the timeless theory of everything), by allowing us to open interesting perspectives towards a unitary and holistic view of physical processes, can be then considered as a fundamental step of this adventure.

Part of the reflection about the science of the last decades has put in evidence the "non-cumulative" aspect in the development of scientific knowledge. According to this view, the evolution of scientific theories is marked by large or small breaking points in which empirical results are reorganized within new theories. These new theories would be consequently, to some extent, "incommensurable" with respect to the previous ones.

The considerations that we have made in this book provide a different key of reading of the evolution of scientific knowledge. In the view analyzed in this book, the fundamental ideas are based on the expectation that quantum mechanics and general relativity represent our best guide to describe physical world in their respective domains. In particular, the timeless theory of everything analysed in this chapter lets us realize that there is a subtle, but very definite cumulative aspect in the progress of physics, which goes far beyond the growth in the validity and precision of the empirical content of the theories. In moving from a theory to another, more general, theory that incorporates it and extends and supersedes it in such a way to include other domains of nature (or in such a way to provide a new key of reading of reality), we do no save only the verified empirical content of the old theory, but there is more. This "more" regards the spectacular an undeniable predictive power of theoretical physics. In fundamental physics the most significant progresses are at the same time linked with intuitions that lead to great conceptual revolutions and with the confidence in the predictive power of the old theories.

The central core of scientific thinking lies in a cautious faith in the known theories and, at the same time, by means of the intuition, in being open to put questions about nature. Exploring the possible ways of thinking about the world, being ready to subvert, if required, our old prejudices, opening the possibility to investigate new scenarios is among the best human adventures. The view of a timeless, granular, wave and vibratory space can be then considered a relevant and significant step of this adventure: it shows how we can have faith in the fundamental theories of physics and, at the same time, how we can modify our view of reality. The scenarios and perspectives, analysed in this book, towards a timeless background as fundamental arena of physics as well as a "timeless theory of everything" can be considered an important step in the process of improvement in our description and understanding of nature.

REFERENCES

[1] Dirac, P. A. M. *Nature* 1951, *168*, 906-907.
[2] Sakharov, A. D. *Doklady Akad. Nauk S.S.S.R.* 1967, *177*, 70-71.
[3] Rugh, S. E.; Zinkernagel, H. *Stud Hist Phil Mod Phys* 2002, *33*, 663-705.
[4] Timashev, S. F. (2011). Physical vacuum as a system manifesting itself on various scales – from nuclear physics to cosmology. arXiv:1107.pdf [gr-qc]
[5] Laszlo, E. *Science and the Akaschic Field*; Inner Traditions: Rochester, NY, 2004.
[6] Laszlo, E. *Risacralizzare il cosmo*; Apogeo/Urra: Milano, 2008.
[7] Wheeler, J. A. In *The Ghost in the Atom*; Davies, P. C. W.; Brown, J. R.; Eds.; Cambridge University Press, Cambridge, MA, 2000.
[8] Clifford, W., cited by Milo Wolf and Geoff Haselhurst in "Einstein's Last Question", VIA: *Jour Integral Thinking for Visionary Action* 2005, *3*, 1.
[9] Einstein, A. *Nature* 1930, *125*, 897-898.
[10] Wolff, M.; Haselhurst, G. *Front Persp* 2004, *13*, 6-15.
[11] Fiscaletti, D.; Sorli, A. *Front Persp* 2005, *14*, 43-54.

[12] Fiscaletti, D.; Sorli, A. *Front Persp* 2006, *15*, 23-28.

[13] Greene, B. *The Elegant Universe: Superstrings, Hidden Dimensions, and the Quest for the Ultimate Theory*; Vintage Books: New York, NY, 1999.

[14] Kaku, M. *Introduction to Superstrings and M-Theory*; Springer: Berlin, DE, 1999.

[15] Rovelli, C. (1997). Loop Quantum Gravity. *Living Reviews in Relativity*, http://relativity.livingreviews.org/Articles/lrr-1998-1/.

[16] Rovelli, C. In *Physics meets philosophy at the Planck scale*; Callender C.; Huggett, N.; Eds.; Cambridge University Press: Cambridge, MA, 2001.

[17] Rovelli, C. *Phys World* 2003, *7*, 1-5.

[18] Rovelli, C. *Quantum gravity*; Cambridge University Press: Cambridge, MA, 2004.

[19] Rovelli, C. (2010). A new look at loop quantum gravity. arXiv:1004.1780v1 [gr-qc]

[20] Garrett Lisi, A. (2007). An exceptionally simple theory of everything. arXiv:0711.0770v1 [hep-th]

[21] Garrett Lisi, A.; Weatherall, J. O. "A geometric theory of everything", *Scientific American*, December 2010, pp. 55-61.

[22] Geroch, R. *General Relativity from A to B*; University of Chicago Press: Chicago, IL, 1978; pp. 20-21.

[23] Smolin, L. (2008). The Problem of Time in Gravity and Cosmology, Lecture 1. http://pirsa.org/pdf/_les/dd4b88c3-7acd-4ea8-91fc-9302bc248e38.pdf

[24] Bohm, D.; Hiley, B. *The Undivided Universe: an Ontological Interpretation of Quantum Theory*: Routledge: London, UK, 1993.

[25] Hiley, B.; Callaghan, R. (2010). The Clifford Algebra approach to Quantum Mechanics A: The Schrödinger and Pauli Particles. arXiv:1011.4031 [math-ph]

[26] Silberstein, M.; Stuckey, W. M.; Cifone, M. *Stud Hist Phil Mod Phys* 2008, *39*, 736-751.

[27] Stuckey, W. M.; Silberstein, M.; Cifone, M. *Found Phys* 2008, *38*, 348-383; e-print arXiv:quant-ph/0510090.

[28] Silberstein, M.; Stuskey, W. M.; McDeVitt, T. (2012). Being, becoming and the undivided universe: a dialogue between Relational Blockworld and the Implicate Order concerning the unification of relativity and quantum theory. arXiv:1108.2261v3 [quant-ph]

[29] Cramer, J. G. *Phys Rev D* 1980, *22*, 362–376.

[30] Cramer, J. G. *Found Phys* 1983, *13*, 887–902.

[31] Cramer, J. G. *Rev Mod Phys* 1986, *58*, 647–88.

[32] Cramer, J. G. *Int Jour Theor Phys* 1988, *27*, 227-236.
[33] Kastner, R. *The new transactional interpretation of quantum theory: the reality of possibility*; Cambridge University Press: Cambridge, MA, 2012.
[34] Fiscaletti, D. *Electr Jour Theor Phys* 2005, *2*, 15-20.
[35] Fiscaletti, D.; Sorli, A. *Scient Inq* 2007, *8*, 65-80.
[36] Fiscaletti, D. *Scient Inq* 2008, *9*, 173-200.
[37] Fiscaletti, D. *IUP Jour Phys* 2010, *3*, 7-28.
[38] Fiscaletti, D. *Prospettive alla ricerca del graal. Verso una visione unitaria di spazio, materia e vita*; Aracne Editrice: Roma, IT, 2010.
[39] Christian, J. (2007). Absolute being vs relative becoming. arXiv:gr-qc/0610049v2
[40] Fiscaletti, D. *Jour Adv Phys* 2012, *1*, 150-160.
[41] Fiscaletti, D.; Sorli, A. "Perspectives about quantum mechanics in a model of a three-dimensional quantum vacuum where time is a mathematical dimension", in preparation, 2014.
[42] Fiscaletti, D.; Sorli, A. "Energy density of space and quantum behaviour", in preparation, 2014.
[43] Albert, D. Z. *Quantum Mechanics and Experience*; Harward University Press: Cambridge, MA, 1992.
[44] Philippidis, C.; Dewdney, C.; Hiley, B. *Nuovo Cimento B* 1979, *52*, 15-28.
[45] Fiscaletti, D. *I fondamenti nella meccanica quantistica. Un'analisi critica dell'interpretazione ortodossa, della teoria di Bohm e della teoria GRW*; CLEUP: Padova, 2003.
[46] Fiscaletti, D. *I gatti di Schrödinger. Meccanica quantistica e visione del mondo*; Muzzio Editore: Rome, 2007.
[47] Dewdney, C. In *Quantum Uncertainties – Recent and Future Experiments and Interpretations*; Honig, W M.; Kraft, D. W.; Panarella, E.; Eds.; NATO ASI Series; Plenum Press: New York, NY; 1987; pp. 19-40.
[48] Bergia, S. In *Quanti Copenaghen? Bohr, Heisenberg e le interpretazioni della meccanica quantistica*: Tassani. I.; Ed.; Il Ponte Vecchio: Cesena, IT, 2004; pp. 179-199.
[49] Wolff, M. *Phys Ess* 1993, *6*, 181-203.
[50] Wolff, M. In *From the Hubble Radius to the Planck Scale*; Amoroso R. L. and others; Eds.; Kluwer Acad. Publ.: Amsterdam, NE, 2002; pp. 517-524.
[51] Wolff, M. (2002). Origin of the Natural Laws in a Binary Universe. www.quantummatter.com/PNASLast.html
[52] Mead, C. *Collective Electrodynamics*; MIT Press: Cambridge, MA, 2000.

INDEX

#

20th century, xi, xii, xiii, xv, 61, 62
21st century, ix, xi, 57

A

abstraction, 3
amplitude, 19, 41, 42, 50
annihilation, 35, 38, 39, 40, 46, 80, 81, 84, 86, 87, 88, 89, 90, 93, 95, 97, 99, 101, 102, 103, 105, 106
Aristotle, 80
arrow of time, 5, 7
ASI, 110
asymmetry, 24
atomic clocks, 51, 52
atomic nucleus, 63
atomism, ix, 61, 75
atoms, 16, 17, 31, 33, 42, 67, 96

B

baryons, 84, 86
Belgium, 58
Big Bang, 54, 55
black hole, 4, 15, 53, 56
bosons, 66, 84, 85, 86, 97
brain, 15

C

capsule, 16
carbon, 96
causality, 33, 92
chaos, 22, 28
Chicago, 56, 74, 109
circulation, 53, 54
classical mechanics, xiii, 1, 18
clock synchronization, 11
closed string, 66
coherence, 40
communication, 17, 20, 21, 25, 27, 28, 31, 32, 91
complementarity, 80
complications, 3
conception, xi, 4, 98
condensation, 6
configuration, 56, 75
conscious experiences, 77
constituents, xv, 36, 47, 63, 65, 67, 81, 105, 106
construction, 37, 46, 52
contingency, 77
contradiction, 15, 52
correlations, ix, 2, 5, 16, 17, 19, 20, 22, 27, 30, 31, 42, 77, 98
cosmos, 75
cycles, 53, 56

D

dark energy, 37
David Bohm, 16, 18
decay, 97
destruction, 37, 38, 39, 41, 42, 43, 44
detection, 38, 41
dichotomy, 77
differential equations, 4
dilation, 12, 13
dimensionality, 37
direct action, 34
directionality, 5
discreteness, 21
dispersion, 82
displacement, 14
disposition, 97
distortions, 13
distribution, 22, 33, 76, 92
distribution function, 76
dualism, 18
dynamic equilibrium of the universe, vi, 53, 54, 56, 81, 96
dynamism, 73, 74, 77

E

electric charge, 62, 72
electromagnetic, 6, 11, 30, 35, 36, 45, 53, 54, 63, 72, 79, 83, 84, 85, 87, 96, 106
electromagnetic waves, 79, 83, 84, 85
electromagnetism, 62, 67, 72, 97
electron, 31, 35, 43, 68, 72, 86, 88, 92, 94, 98, 102
electrons, 31, 43, 67, 72, 84, 92, 93, 96
elementary particle, 17, 18, 22, 28, 29, 34, 39, 43, 47, 59, 70, 71, 72, 90
emission, 35, 42, 78
emitters, 79
encoding, 33
energy, xiii, 18, 22, 28, 35, 36, 37, 38, 39, 40, 41, 42, 43, 46, 47, 48, 49, 50, 53, 54, 55, 56, 61, 62, 63, 64, 65, 69, 75, 78, 79, 81, 82, 83, 84, 85, 86, 87, 88, 89, 90, 91, 93, 94, 95, 96, 97, 98, 99, 100, 101, 102
energy conservation, 95
energy density, 35, 36, 37, 38, 39, 40, 41, 46, 47, 48, 49, 50, 53, 54, 62, 63, 81, 82, 83, 84, 85, 86, 88, 89, 91, 93, 96, 99, 100, 101
energy transfer, 35, 90, 93, 95, 97
entropy, 5, 22, 28, 55, 82, 83, 86, 102
EPR, v, 16, 17, 18, 20, 22, 27, 28, 30, 31, 34, 77, 78, 91
EPR experiment, 16, 17, 27
equilibrium, xi, 2, 53, 54, 55, 56, 81, 84, 96
Euclidean space, 6
evolution, xii, 1, 2, 4, 5, 16, 22, 37, 38, 40, 43, 44, 56, 71, 74, 75, 80, 96, 105, 107
excitation, 33, 36

F

faith, 107, 108
fermions, 73, 84
fiber, 72
fibers, 71, 72, 73
field theory, 63, 67, 69, 71, 98
filament, 67
fluctuations, 35, 45, 62, 63
force, 6, 18, 19, 36, 43, 47, 48, 50, 62, 68, 72, 85, 90, 96, 97, 98, 103
formation, 56, 96
formula, 32
foundations, 81, 101

G

galaxies, 53, 54, 55
general relativity, xi, xii, xiii, xiv, xv, 3, 5, 15, 37, 45, 51, 52, 61, 64, 67, 68, 76, 77, 81, 107, 109
geometry, xii, 5, 28, 45, 50, 51, 58, 60, 62, 65, 66, 69, 71, 73, 75, 76
gluons, 63, 67, 85
God, 54, 98
GPS, 51

gravitation, 5, 8, 30, 45, 46, 48, 50, 67, 68, 97, 98
gravitational constant, xv, 37
gravitational effect, 49, 51, 52
gravitational field, xii, xiv, 45, 47, 66, 67, 69, 70, 103
gravitational force, 47
gravity, ix, xii, xiii, xiv, xv, 2, 4, 5, 30, 31, 41, 45, 46, 47, 48, 49, 50, 51, 52, 55, 57, 59, 61, 66, 67, 69, 70, 71, 72, 73, 74, 76, 84, 87, 103, 104, 105, 109
growth, 107

H

Hamiltonian, 4, 6, 82
helium, 32
Higgs boson, 73
Higgs field, 63
Higgs particle, 46
Hilbert space, 33, 41
holistic view, 107
human, 14, 105, 107, 108
hydrogen, 36, 69, 88

I

illusion(s), 32, 75, 77, 80
image, 66, 100
immediate symmetric interpretation of quantum non-locality, 33
incompatibility, 67, 80
independence, 67, 69, 106
independent variable, 1
inequality, 20
inertia, 4, 59
information exchange, 77
initial state, 38
interference, 76, 92
internal time, xii
ionization, 32

J

Josephson junction, 36

L

Large Hadron Collider, 71
laws, xi, xv, 6, 12, 23, 24, 35, 40, 43, 75
lead, 12, 21, 50, 96, 108
light, xi, 1, 9, 11, 13, 16, 30, 32, 36, 45, 48, 51, 52, 55, 62, 64, 72, 79, 81, 83, 91, 92, 95, 104
loop quantum gravity, ix, xv, 8, 35, 61, 66, 68, 69, 70, 71, 76, 81, 100, 103, 104, 105, 109

M

magnetic field, 72
magnitude, 36, 67
mapping, 105
Maryland, 69, 78
mass, 26, 37, 45, 46, 48, 52, 53, 56, 62, 63, 68, 83, 85, 86, 100
massive particles, 52
material bodies, 65
matter, ix, xi, xii, xiii, xiv, 1, 2, 9, 21, 25, 35, 38, 49, 50, 51, 53, 54, 55, 56, 58, 61, 63, 64, 65, 67, 68, 71, 75, 76, 77, 79, 80, 81, 82, 83, 84, 85, 87, 96, 98, 99, 100, 101, 103, 104
measurement, xiv, 1, 2, 6, 7, 9, 13, 17, 20, 24, 26, 37, 69, 85
mesons, 84, 86
Milky Way, 56
modelling, ix, 61, 73
models, ix, xi, xiii, 3, 25, 37, 46, 59, 71
modifications, 84, 87, 97
modulus, 19, 92
molecular structure, 96
molecules, 96
momentum, xiii, 32, 62, 70, 73

N

NATO, 110
nature of time, 57, 77
neutrinos, 67
neutron stars, 56
neutrons, 63, 67
Newtonian physics, xi, 1, 9, 23
nodes, 69, 70
non-Euclidean geometry, 5
nuclei, 96
nucleons, 36
nucleus, 36, 56
numerical order of material change, ix, xiv, 1, 2, 3, 7, 9, 10, 12, 13, 14, 15, 16, 17, 24, 25, 27, 28, 29, 40, 56, 61, 80, 84, 102, 104

O

orbit, 51
organize, 107
organs, 79
oscillation, 32

P

paradigm shift, xii
parallel, 10, 11, 13
particle physics, xiii, 62, 64
phase transformation, 7, 85
photons, 33, 36, 63
physical fields, 34, 84, 87, 96, 97
physical laws, 4, 6, 24, 25
physical phenomena, ix, 29, 30, 31, 61, 64
physical properties, 38, 46, 65
physical structure, xiv, 72, 104
physical theories, xi, xiii, xv, 9, 20, 25, 75, 81
physics, ix, xi, xii, xiii, xiv, xv, 1, 2, 3, 4, 5, 6, 7, 10, 18, 34, 37, 50, 58, 61, 62, 63, 64, 65, 66, 67, 68, 69, 71, 72, 73, 74, 75, 76, 77, 80, 81, 82, 84, 87, 91, 98, 105, 107, 108
Planck constant, xiii, 18, 26, 37
planets, 13, 47
polarization, 32, 35, 62
preparation, 38, 41, 59, 110
principles, xiv, 6
probability, 33, 39, 50, 69, 92, 94, 95
propagation, 41, 43, 44, 45, 48
proposition, 52
protons, 63, 67

Q

quanta, xv, 34, 35, 40, 43, 45, 46, 53, 56, 64, 65, 66, 69, 70, 78, 79, 80, 81, 82, 83, 84, 85, 86, 87, 88, 89, 90, 93, 95, 96, 97, 98, 99, 100, 101, 102, 105, 106
quantization, 8, 74
quantum chromodynamics, 36
quantum cosmology, xiii
quantum dynamics, 41
quantum electrodynamics, 62, 81
quantum entanglement, 31, 77
quantum field theory, 35, 36, 64, 66, 69, 71
quantum fluctuations, 36
quantum foam, 75, 98
quantum gravity, ix, xiv, xv, 4, 8, 35, 45, 61, 66, 69, 70, 71, 73, 74, 76, 81, 100, 103, 104, 105, 109
quantum mechanics, xiii, xiv, xv, 1, 4, 9, 18, 19, 20, 23, 24, 25, 26, 33, 35, 37, 39, 41, 43, 59, 67, 68, 69, 70, 75, 76, 78, 79, 91, 92, 94, 99, 107, 110
quantum optics, 32
quantum phenomena, 20, 25, 27, 29, 49, 66, 77, 91
quantum potential, 18, 19, 21, 22, 24, 25, 26, 27, 28, 29, 31, 32, 33, 34, 38, 39, 40, 43, 44, 49, 50, 65, 75, 88, 89, 90, 91, 92, 93, 95, 97, 100, 102
quantum realm, 66
quantum state, 32, 87
quantum theory, xi, xii, xiii, 4, 5, 17, 19, 20, 35, 45, 57, 58, 60, 61, 66, 67, 69, 75, 76, 77, 87, 88, 109, 110
quarks, 36, 63, 67, 84, 86, 96

R

race, 14
radar, 13
radiation, xiii, 11, 33, 35, 82
reading, x, 22, 41, 70, 80, 94, 95, 102, 103, 107
real time, 32, 61
reality, ix, xiii, xiv, xv, 2, 3, 4, 5, 7, 9, 10, 15, 16, 17, 20, 25, 27, 34, 38, 40, 41, 42, 43, 44, 50, 54, 61, 63, 64, 65, 73, 79, 80, 81, 82, 83, 90, 91, 93, 95, 96, 102, 103, 104, 107, 108, 110
red shift, 52, 55
reference frame, 4, 11, 13, 29, 30, 49
reference system, 1, 29
relativity, xi, xii, xiii, xiv, xv, 3, 5, 9, 10, 11, 12, 13, 15, 30, 31, 37, 45, 47, 51, 52, 54, 61, 64, 66, 67, 68, 73, 74, 76, 77, 81, 82, 107, 109
rotations, 72

S

scattering, 66
school, 49
Schrödinger equation, 1, 22, 24, 26, 27, 39, 41, 69
scientific knowledge, 107
senses, 82, 83, 86, 93, 100
shape, 47, 72
showing, xiii
signals, 11, 33, 79
Singapore, 58
solution, 26, 75
spacetime, 57, 78, 79, 80
space-time, xii, xiv, 3, 5, 8, 9, 12, 20, 21, 34, 35, 37, 38, 39, 40, 44, 45, 47, 48, 49, 50, 51, 54, 57, 61, 63, 63, 64, 66, 69, 71, 72, 73, 74, 75, 76, 77, 78, 79, 82, 98, 102, 104
spatial information, 19
special relativity, ix, xi, xiii, 2, 5, 9, 10, 11, 12, 13, 30, 52, 74, 82
special theory of relativity, 9, 11, 12, 15
species, 63
speed of light, xv, 8, 13, 37, 62, 64, 84
spin, 16, 18, 62, 70, 71, 72, 84, 88, 103
Standard Model, 62, 63, 64, 71, 72
stars, 56
state, 1, 5, 16, 17, 21, 24, 26, 27, 32, 33, 34, 37, 38, 39, 41, 44, 46, 55, 70, 72, 82, 83, 86, 87, 88, 90, 91, 93, 95, 97, 101, 102
states, 3, 5, 17, 19, 34, 36, 46, 70, 71, 73, 81, 84, 87, 96, 97, 99, 101, 102, 103
statistics, 38, 84
storage, 22
string theory, xv, 72, 75, 76, 104
strong force, 62, 72
strong interaction, 36, 63, 66, 85
structure, xi, xiv, xv, 8, 10, 21, 22, 35, 36, 42, 45, 54, 63, 65, 67, 69, 71, 73, 76, 78, 82, 84, 86, 87, 88, 90, 93, 98, 100, 103, 104, 107
structuring, 54
succession, 5, 96
superfluid, 45
superstring theory, ix, 61, 66, 67, 68, 81, 102, 103, 104, 105
surface region, 36
symmetry, 6, 24, 25, 26, 27, 29, 34, 72, 73, 85
synchronization, 11, 13
synthesis, xiv, xv, 12, 18, 22, 23, 29, 34, 50, 68, 70, 71, 83, 85, 87

T

temperature, 62
tempo, 57
tension, 74, 75
tensions, 77
Theory of Everything, 80, 98
thermodynamics, 53, 83
three-dimensional space, 2, 10, 12, 15, 17, 19, 21, 22, 27, 28, 31, 35, 81, 84
three-dimensional timeless quantum vacuum, 41, 50
throws, 16
time, ix, xi, xii, xiv, xv, 1, 2, 3, 4, 5, 6, 7, 8, 9, 10, 11, 12, 13, 14, 15, 16, 17, 20, 21, 23,

24, 25, 26, 27, 28, 29, 30, 31, 32, 33, 34, 35, 37, 38, 39, 40, 41, 42, 43, 44, 45, 47, 48, 49, 50, 51, 52, 54, 56, 57, 59, 60, 61, 62, 63, 64, 65, 66, 68, 69, 70, 71, 72, 73, 74, 75, 76, 77, 78, 79, 80, 81, 82, 84, 87, 88, 91, 92, 94, 96, 97, 98, 100, 101, 102, 103, 104, 106, 107, 108, 110
time variables, 11
timeless theory of everything, x, 61, 81, 88, 91, 96, 99, 100, 103, 104, 105, 107, 108
total energy, 23
trajectory, 48, 88, 89, 90, 92, 93, 94, 95, 98, 100, 102
transactions, 37, 38, 78, 79, 80
transformations, 6, 7, 10, 11, 13, 37, 55, 81, 85
translation, 28, 56
transmission, 32, 98
treatment, 53, 67, 68, 72, 104
tunneling, 31, 32

U

UK, 57, 58, 59, 60, 109
universe, ix, xi, xii, xiv, xv, 1, 2, 3, 4, 5, 7, 9, 10, 12, 14, 15, 16, 21, 24, 29, 35, 36, 37, 40, 52, 53, 54, 55, 56, 58, 62, 63, 64, 65, 66, 67, 68, 71, 73, 76, 77, 78, 79, 80, 81, 82, 83, 84, 85, 96, 99, 100, 101, 102, 104, 105, 106, 109

V

vacuum, ix, 2, 13, 22, 28, 35, 36, 37, 38, 39, 40, 41, 42, 43, 44, 45, 46, 47, 48, 49, 50, 53, 54, 59, 61, 62, 63, 64, 78, 80, 81, 82, 83, 84, 85, 86, 87, 88, 89, 90, 91, 93, 94, 95, 96, 97, 99, 100, 101, 102, 103, 104, 105, 106, 108, 110
variables, xii, 17, 33, 39, 85
variations, 65
vector, 32, 39
velocity, 11, 12, 13, 14, 15, 30, 33, 36, 51, 52
vibration, 21, 68, 81, 83, 85, 86, 89, 90, 93, 97, 99, 100, 101, 102, 103, 105, 106

W

weak interaction, 85
Wheeler-DeWitt equation, 69, 70, 75